BYPASSING

BYPASS

SURGERY

Books by Elmer M. Cranton, M.D.

Trace Elements and Nutrition
(with Richard Passwater, Ph.D.)

Bypassing Bypass

A Textbook on EDTA Chelation Therapy
(editor; foreword by Linus Pauling, Ph.D.)

Bypassing Bypass (2nd ed.)

Resetting the Clock: Five Antiaging Hormones That Are Revolutionizing the Quality and Length of Life
(with William Fryer)

A Textbook on EDTA Chelation Therapy
(editor; 2nd ed.)

BYPASSING BYPASS SURGERY

ELMER M. CRANTON, M.D.

HAMPTON ROADS
PUBLISHING COMPANY, INC.

Cover design by Marjoram Productions

Medex Publishers
503 First Street South
P.O. Box 1000
Yelm, WA 98597-1000
360-458-1077
fax: 360-458-1661
drcranton@drcranton.com
www.drcranton.com

For information on distribution:
Hampton Roads Publishing Company, Inc.
1125 Stoney Ridge Road
Charlottesville, VA 22902

434-296-2772
fax: 434-296-5096
e-mail: hrpc@hrpub.com
www.hrpub.com

If you are unable to order this book from your local
bookseller, you may order directly from the publisher.
360-458-1077
fax: 360-458-1661
1-800-742-5682, toll-free.
Library of Congress Catalog Card Number: 2001092122
ISBN 1-57174-297-2
10 9 8 7 6 5 4 3 2
Printed on acid-free paper in Canada

Dedicated to

Norman E. Clarke Sr., M.D. (1892–1984),

originator of EDTA chelation therapy to treat atherosclerosis. Dr. Clarke, an eminent cardiologist and chief of research at the Providence Hospital in Detroit, was first to hypothesize that EDTA might benefit patients with heart disease. He designed and conducted the first clinical trials and published the first research data on this subject. For two decades he was the principal spokesman who kept this therapy active and under study.

DISCLAIMER

This book is educational in nature and is not to be used as a basis for medical diagnosis or treatment. The information contained herein is based on the many years of experience of Elmer M. Cranton, M.D., and on an extensive review of the scientific literature. However, this book is definitely not intended to substitute for medical care by a licensed health care professional. Treatment of individual health problems must be closely supervised by a physician. The author recommends against changing any current medication, altering current therapy, or adding any new therapies without first consulting a physician. Chelation therapy is a controversial subject and widely conflicting opinions on the contents of this book are held by various factions within the medical profession.

TABLE OF CONTENTS

BYPASS SURGEONS AND MEDICAL SCHOOL PROFESSORS ENDORSE CHELATION THERAPY

The following are five letters of endorsement for EDTA chelation therapy written to me in response to an earlier version of this book. I was delighted to receive such favorable backing from three bypass surgeons and two other medical school professors. These letters are reprinted here as evidence that EDTA chelation therapy is gaining acceptance in academic medicine. All five doctors offer EDTA chelation therapy to their patients.

"In his book, *Bypassing Bypass*, Dr. Cranton very clearly explains new concepts relative to the aging process and atherosclerosis. As a practicing cardiovascular surgeon, I and many of my associates have patients who are not surgical candidates. These patients are then often relegated to a life of continued disability and pain. A member of my family fell into this group and was told to 'go to a nursing home and die.' He was instead treated with EDTA chelation therapy and is alive and comfortable three years later. *I often observe similar benefits for patients in my own practice who have had chelation therapy.* Those of us in academic medicine and surgery should put aside our blinders, open our minds, and delve further into any promise of improvement for those unfortunates who have no other hope. *Everyone interested in the betterment of life should read Dr. Cranton's book."*

Ralph Lev, M.D., M.S.
Clinical Associate Professor of Surgery
Vascular Surgeon and Chief of
Cardiovascular Surgery
John F. Kennedy Medical Center
New Jersey Medical School

"Since World War II basic scientists and some clinicians have been increasingly aware of the critical role that many metals play in normal and disease states. Only recently has the biologic significance of cross-linking and free radicals been recognized. Chelation favorably influences the role of all of these. *This book, the first to bring together for discussion metals, cross-linking, and free radicals, may well be the necessary impetus to the inevitable general acceptance of chelation therapy."*

John H. Olwin, M.D.
Vascular Surgeon and
Clinical Professor of Surgery, Emeritus
Rush Medical College and
University of Illinois
Attending Surgeon, Emeritus, Rush
Presbyterian St. Luke's Medical Center

"*Bypassing Bypass* is a very elucidating book, explaining the often misunderstood effects of EDTA chelation therapy. I only wish that I had been familiar with this therapy when I first began my career as a cardiovascular surgeon. I would have been more selective in my choice of patient for bypass surgery. *I now achieve more lasting results with less risk, enhancing the benefits of surgery, and often avoiding surgery, by providing chelation therapy for my patients.*"

Peter J. van der Schaar, M.D., Ph.D.
Cardiac Surgeon and Director
International Biomedical Center
Netherlands

"In this book Dr. Cranton has taken a quantum step forward in the task of reconciling chelation therapy to traditional medical thinking and understanding. By a series of logical steps, he shows the reader the meaning behind the most recent discoveries concerning all forms of degenerative diseases.

"Specifically, Dr. Cranton explains the free radical theory of degenerative disease. This new and exciting explanation, which builds upon facts first uncovered in the 1960s, suggests that underlying most degenerative disease from heart attacks to cancer is excess free radical activity. If this is the unifying factor behind mankind's major twentieth-century health hazards, then this book will perform the service not merely of demonstrating the merits of an invaluable health care treatment, chelation therapy, but of pinpointing what may be one of the major medical breakthroughs of the century.

"For the patient who already has atherosclerosis, the prospect of reversing his disease through chelation treatments is welcome news. It has been shown in experimental animals as well as by blood flow studies in humans that the deposits in arterial walls are reversible.

For those people who have not yet begun to feel the effects of atherosclerosis, this book brings home the lessons of preventive medicine. Put very simply, it is wiser to keep one's body in a state of health than to let it deteriorate to a point where a risky surgical procedure such as bypass heart surgery is found advisable.

"*Bypassing Bypass* is a book that will help patients to take responsibility for their own health, and it must be considered required reading for every serious student of preventive medicine physician and patient alike."

H. Richard Casdorph, M.D., Ph.D.
Assistant Professor of Clinical Medicine
University of California, Irvine
Long Beach, California

"Dr. Elmer Cranton, a graduate of Harvard Medical School, class of 1964, is living proof that you don't have to be on the faculty of a medical school and do basic research to be 'academic.'

"Dr. Cranton, apparently driven by the desire to understand how and why EDTA works, i.e., alleviates the pain of angina and intermittent claudication and decreases symptoms of shortness of breath and fatigue in patients with coronary artery disease and peripheral vascular disease, has come up with a plausible hypothesis based on the theory of 'Free Radical Damage and Lipid Peroxidation of Cell Membranes.'

"What this amounts to is an explanation which is either 'The Greatest Story Ever Told' or a reasonable explanation of some very complex phenomena, related to aging and the degenerative diseases. My hunch is that it is the latter.

"We can also thank God that some people do not fit into the usual cookie mold and they really do march to a different drummer. And also that there are still some physicians who rely heavily on their own clinical experience, as to what works and what doesn't work, rather than on what has become known as usual and customary in modern medical practice."

James P. Carter, M.D., Dr. P.H.
Chairman and Professor,
Department of Nutrition
Tulane University School of Public Health and
Tropical Medicine
New Orleans, Louisiana

FOREWORD

In the mid-1970s I became discouraged with my career in medicine. My patients were growing older, and traditional treatments for degenerative diseases of aging did not slow progression of the underlying disease process. Little emphasis was given to preventive medicine.

Patients with cardiovascular diseases caused by atherosclerotic plaque within arteries, starving vital organs of blood flow, were especially frustrating to see. Traditional therapies for angina, heart attack, stroke, senility, and gangrene leading to amputation of legs brought only partial relief of symptoms when successful and did not slow or reverse the underlying disease process. It was a stopgap approach, aimed principally at symptom relief rather than disease reversal.

Early excitement over the advent of bypass surgery and angioplasty was not justified. Clinical results of invasive and surgical procedures were too often temporary, with a high incidence of alarming complications and death. Following bypass, the underlying disease process soon emerged once more.

When I first learned about EDTA chelation therapy and its associated program of preventive medicine, I discovered an approach to health care that really excited me. My subsequent experiences were very similar to those of Dr. Elmer Cranton, as he relates in this book.

Like Dr. Cranton, I studied extensively in the scientific literature and then introduced those innovative principles to my patients. The remarkable improvements I observed in my own practice were amazing and gratifying to me and were just like those described by Dr. Cranton. If anything, Dr. Cranton has understated the benefits of EDTA chelation therapy.

The major emphasis of my medical practice is now on early detection and prevention of disease—including lifestyle improvements to reduce risk factors, proper nutrition, and nutritional supplementation—to form a comprehensive program of health promotion and preventive medicine. EDTA chelation therapy is an important part of my practice.

I routinely observe reversal of underlying cardiovascular disease in my patients using EDTA chelation therapy, without the risk and expense of surgery and other invasive therapies. The patients' inherent healing mechanisms are invoked. This type of therapy often eliminates the need for vascular surgery or angioplasty. There is less reliance on prescription medications, with their many potential side effects and high cost.

My medical practice has become much more fulfilling and pleasurable—both for me and for my patients. But, my practice and Dr. Cranton's practice are not unique. Neither are the benefits experienced by our patients. An increasing number of physicians administer this therapy, and a million or more of their patients have now undergone EDTA chelation—almost as many as have had bypass surgery and angioplasty. A very large majority of chelation patients experience the same remarkable benefits described by Dr. Cranton in the following pages.

Then why is EDTA chelation therapy not yet an integral part of standard medical practice? That's a good question with no good answer. Dr. Cranton portrays for you a politically powerful medical and economic system, which continues to delay and suppress this therapy's widespread acceptance.

This outstanding book accurately describes a complex treatment program in an easily understandable manner, and gives an extensively referenced scientific basis for its action. To make informed decisions about your health, read this book! You too may bypass that bypass, slow the aging process, and improve the quality of your life.

James P. Frackelton, M.D.
Past President, American College for
Advancement in Medicine
Preventive Medicine Group
24700 Center Ridge Road
Cleveland, OH 44145
(440) 835-0104

PREFACE

This book tells you about a simple, nonsurgical treatment, administered in a doctor's office, that in the majority of cases does away with need for bypass surgery or angioplasty.

This edition has been extensively revised and updated, with much new material added since the original versions—which sold close to two hundred thousand copies over sixteen years. Every word has been reviewed and updated. Several new chapters have been added. More than half of this current volume is new material. Chapters on chelation research, bypass surgery, angioplasty, diet, and the scientific rationale are largely new. New chelation research and long-term follow-up studies of bypass surgery and angioplasty are summarized. New case histories are scattered throughout, and many more are added in an appendix. Changes and additions are so extensive, that I consider this to be largely a new book, built around the skeleton of the original volume.

Cardiovascular disease causes half of all death and disability in the United States. At any one time, more than forty million men and women suffer with symptoms of heart disease, America's number one killer. Coronary heart ailments continue at epidemic proportions—one million people die each year. Many are struck down without warning; others succumb after years of painful, debilitating angina—a sure sign that coronary arteries are not delivering enough blood to some area of the heart.

Angina (commonly pronounced "*ann*-gin-ah") means pain and *pectoris* means chest. Angina pectoris (usually shortened to angina) is medical jargon for "chest pain" and refers to pain radiating from the heart when blood flow in coronary arteries is blocked by atherosclerotic plaque.

Coronary artery bypass surgery or grafting (commonly abbreviated as CABG and referred to informally as "cabbage" by doctors) is a procedure in which blocked portions of major coronary arteries are bypassed with grafts from a patient's leg veins. Percutaneous coronary balloon angioplasty (abbreviated as PTCA), with or without insertion of stents, is equally popular. These procedures are currently the most frequent surgical solutions to the nation's leading medical problem. Over the past three decades, since first coming into widespread use, both the frequency and the cost of heart bypass operations have escalated. Since 1968, more than five million heart patients have undergone this surgical treatment. Bypass surgery has now grown to a $25-billion-a-year industry. The cost for balloon angioplasties and stents doubles that figure to $50 billion. (A more detailed analysis will be found in chapter 16.)

Reasonable men and women might logically assume that the runaway popularity of these enthusiastically prescribed treatments is based on convincing and demonstrable evidence that they prolong life—or, at least, have the potential to prevent further health deterioration. If that were indeed true, there would be far less need for me to write, or you to read, this book. Chapter 16 gives you the full facts and figures on bypass and other invasive treatments. They are not nearly as safe or effective as you might believe.

Don't get me wrong. I'm not totally opposed to bypass surgery or angioplasty. But I am opposed to immediate resort to those risky and expensive procedures without first trying a much safer, simple, inexpensive, and noninvasive office procedure—EDTA chelation therapy. Most patients would not consider surgery after receiving chelation.

I refer patients for surgery when they are unstable, worsening rapidly, and at immediate risk for heart attack or stroke, or when chelation fails, as it sometimes does. If they successfully survive their surgery or angioplasty, I still recommend chelation to prevent recurrence and to avoid further surgery in the future.

How worthwhile is a piecemeal surgical treatment that detours only a few impaired arteries, while many others throughout the body continue to deteriorate? There is an inherent fallacy in bypassing only one or even several restricted portions of the body's blood vessels when the same degenerating condition affects many other segments throughout the entire cardiovascular system. The bypass approach treats the tip of the iceberg—the sites where plaque has developed most rapidly—while ignoring the rest of the circulatory network. At best, it's an expensive stopgap measure, a risky, high-priced surgical "aspirin," providing pain relief and not much more.

The fact that many bypass patients do experience a reduction of pain and remain angina-free for two to five years may be as much a minus as a plus since the absence of chest pain may encourage them to believe themselves medically cured. The patient who thinks himself restored to health, when he is not, may well return to his former unhealthy lifestyle, not realizing he is still in jeopardy.

With all the publicity given to its risks and drawbacks, why do so many continue to opt for the bypass operation or angioplasty? The answer: Atherosclerotic patients have been misled into expecting better results than objective statistics warrant, and they are rarely offered alternatives such as chelation therapy.

"What choice did I have?" is a common patient question. "My doctor had nothing else to suggest."

Arterial dilating drugs such as nitroglycerin help to relieve angina. Blood thinners can cause as big a problem as they correct, but are sometimes indicated. The newer medications such as beta-blockers and calcium antagonists relieve symptoms but have not proved universally useful. And the most worthy alternative—EDTA chelation therapy—is almost never mentioned. Patients are not given a choice. I believe that's wrong.

Chelation therapy is a nonsurgical medical treatment that improves metabolic and circulatory function in many different ways by rebalancing and removing metal ions in the body. This is accomplished by administering a synthetic amino acid, ethylenediaminetetraacetate (EDTA), by an intravenous infusion using a tiny, twenty-five-gauge needle or Teflon catheter. This is done in a doctor's office, without need for expensive hospital care or teams of high-tech specialists.

An informal survey of well-read followers of the latest health news reveals that few had heard of chelation therapy; fewer still had more than a vague idea of what it is. In those rare instances where an individual knew that chelation therapy offers a safe and effective alternative to bypass surgery and angioplasty, he or she had "stumbled" across the information accidentally—a friend or relative had been successfully chelated and had passed the word along. Chelation therapy may be one of the best-kept medical secrets. In recent years almost as many patients are being chelated as bypassed, in most cases with good results—and with a forty-year record of safety.

Thus the real reason for this book extends far beyond the overuse and excessive reliance on bypass surgery, angioplasty, and stents. The fact is that a major therapy (EDTA chelation), which offers a greatly improved quality of life for millions of people with many different age-related diseases, has been largely overlooked.

The disservice to the public cannot be overestimated. Chelation therapy has shown itself to be of value in a variety of supposedly incurable diseases in a significant percentage of patients. While it is true that physicians who practice chelation have not had the tens of millions of dollars necessary to fund a study meeting FDA standards for label claims, that fact does not discredit chelation. Chelating physicians are private practitioners with limited resources, unable to fund the large-scale, well-designed, and scientifically irrefutable double-blind studies that impress the medical community. Properly done, such research would without doubt confirm many smaller studies that have demonstrated the extraordinary value of this treatment. All clinical trials to date have shown benefit (as discussed in chapter 10).

Why has funding for chelation research been so hard to come by? Perhaps the pervasive antichelation stance is easier to understand with the knowledge that chelation therapy does not require the vast resources of modern medical centers. It makes no use of the full panoply of space-age technology so prominent at research-oriented hospitals. While there is nothing about chelation that smacks of wizardry or quackery (it cannot be administered by anyone other than a fully licensed medical doctor), it can be done in a clinician's office on an outpatient basis. It is totally legal for any licensed physician to administer this therapy. This is a benefit to patients but a decided drawback to funding. There can be no profit in proving the validity of chelation therapy to those who most influence funding decisions.

Chelation therapy is nonsurgical and requires only thirty or so visits (lasting several hours each) to a physician's office for an intravenous infusion of an FDA-approved medicine, the organic amino acid EDTA (ethylenediaminetetraacetic acid). EDTA interrupts and allows reversal of the disease process in ways I will explain later. The effects of treatment can be dramatic. Often patients who could not walk across the floor without taking nitroglycerin for their angina pains have recovered and are out playing golf in a matter of a few months.

The seemingly miraculous, "too-good-to-be-true" flavor of many of the testimonials to chelation therapy's benefits has hindered, rather than helped, its acceptance. No wonder! Medical scientists have only recently begun to comprehend the interrelationship of many apparently unconnected ailments.

Chelating physicians have been discomfited, rather than pleased, to have patients report improvement of symptoms of a whole host of diseases, including strokes, angina, heart failure, arthritis, scleroderma, macular degeneration, and diabetic complications, when they could not explain the reasons behind their recovery. Such benefits have been considered so outlandishly improbable by mainstream medicine that rumors regarding them have only served to cast further doubts on the credibility of chelation's advocates.

But science is beginning to make good sense out of what had been considered a bad joke. The very latest buzzwords in scientific circles are "free radical pathology." Scientists in laboratories in all parts of the world are agog over the discovery that there is a common denominator among many degenerative diseases—atherosclerosis included—and that this common denominator is a disease mechanism called free radical pathology. This is a process caused by production in our bodies of a form of oxygen that, in its unstable state, is called superoxide and hydroxyl radicals, peroxide, and singlet oxygen. You'll be reading more about this in later chapters.

Oxygen sometimes "goes wrong" in the body at certain points and under certain conditions, and this excited and damaging form of oxygen reacts with literally anything nearby, causing injury to tissues and cells that is akin to radiation exposure.

In contrast to surgical and other medical modalities, chelation therapy counteracts the underlying disease process. Once intravenously injected into the bloodstream, EDTA can remove causes of excess free radical production, protecting the tissues and organs from further damage. Over time, these injections slow or halt the progress of the free radical damage—an important underlying condition triggering the development of atherosclerosis and many other age-related diseases. This gives the body time to heal and allows restoration of blood flow through occluded arteries—relieving symptoms of arterial insufficiency in every part of the body. Chelation rebalances a variety of metal ions within the body, including nutritional trace elements, improving health and metabolism in ways we are just now coming to appreciate.

Scientists who once scoffed at the therapeutic benefits of nutrients such as vitamin C, vitamin E, selenium, manganese, zinc, beta-carotene, etc. are apt to rethink their position in light of the proven biological antioxidant properties of these substances. Increasingly more studies have been published to document the effectiveness of antioxidant strategies. EDTA chelation is just such a therapy.

Unlike the surgical and invasive approach that in effect assumes that vascular disease is a localized ailment (which it is not), chelation therapy addresses that fact that the condition affects not only individual arteries, such as the coronaries around the heart, but also the arteries to many other organs in the body, and even the tiniest arterioles and capillaries in toes, fingers, and brain.

More than sixty thousand people per year lose their legs from gangrene caused by arterial blockages. Many more than that have strokes. Bypass surgeries to arteries in the legs, neck, and even inside the skull are increasingly common. Chelation would work as well or better in many of those patients, with virtually no danger and much less expense.

Even if you are at this moment symptom-free with no indication that atherosclerotic plaque is insidiously accumulating in your arteries, gradually reducing life-preserving circulation, you nonetheless suffer some degree of free radical pathology, which is inevitable with aging. A slow breakdown of biochemical efficiency inexorably leads to interference with the body's structure and functions. This process starts in early adulthood and becomes increasingly debilitating in later life, varying with inherited resiliency, environmental exposure, use of tobacco, diet, nutritional supplementation, and lifestyle.

Unless you embrace a program of health-promoting strategies designed to prevent biological deterioration, you are sure to develop some form of free radical disease—diabetes, arthritis, Parkinson's, senility, Alzheimer's, atherosclerosis, or cancer, to name just a few of the ailments mistakenly thought to be the inescapable consequences of so-called "normal aging." By the age of eighty-five, half of all Americans have symptoms of Alzheimer's syndrome. Preventive medicine is far preferable to crisis intervention, which often occurs too late to prevent damage. When the breakdown comes, you may well find yourself in a crisis-care predicament, faced with a treatment-decision dilemma. Under the urgency of the moment, chances are slim that you would be offered the chelation option.

Prior to reading this book, you would have had to be extremely lucky to know such an option even existed. That is not apt to be true much longer. There are now a large number of people who have benefited from chelation. They spread the word. A groundswell of support for this therapy is emerging from the lay public, rather than from university medical centers.

Chelation is a remarkable, restorative, life-prolonging treatment that reverses and prevents the symptoms of not only atherosclerosis but also many other age-related degenerative diseases.

Periodic updates and related material will be posted on my website: www.drcranton.com

Elmer M. Cranton, M.D.
Mount Rainier Clinic
503 First Street South, Suite 1
P.O. Box 5100
Yelm, WA 98597-5100
Phone: (360) 458-1061
Fax: (360) 458-1661
e-mail:
drcranton@drcranton.com

Elmer M. Cranton, M.D.,
and Eduardo Castro, M.D.
Mount Rogers Clinic
799 Ripshin Road
P.O. Box 44
Trout Dale, VA 24378-0044
Phone: (540) 677-3631
Fax: (540) 677-3843

1

CHELATION? IT MUST BE SOMETHING NEW!

"Chelation? It must be something new."
Why do you say that?
"I never heard of it."
So?
"It couldn't be any good."
Why?
"Good news spreads fast!"

"Chelation? That doesn't work."
Why do you say that?
"It's been around for years."
So?
"It couldn't be any good."
Why?
"It would be popular by now!"

Objections to chelation on either basis are unfounded. The truth is that chelation therapy is both old and new.

The history of chelation therapy can be traced back to 1893 and the pioneering research of Swiss Nobel laureate

Alfred Werner, who developed the theories that later became the foundation of modern chelation chemistry. Werner's concept on how metals bind to organic molecules opened the new field of chelation chemistry.

It was not until the early 1920s that "chelation" was introduced as an industrial tool, finding wide application in the manufacture of paint, rubber, and petroleum. Chelation was also found to be useful in the separation of specific metals, and it gained importance in electroplating and industrial dye manufacture.

In the mid-1930s, German industrialists, concerned with their reliance on imports from potentially unfriendly countries, embarked on a major effort to develop their own chelating agents. Citric acid, a chelating compound used for stain prevention in the printing of textiles, was high on their substitution list.

Their mission: to develop a unique and patentable calcium-binding additive to keeps stains from forming when calcium in hard water reacted with certain dyes. The substance developed—ethylenediaminetetraacetate, or EDTA, as it is commonly known—was patented in 1935. EDTA proved to be more successful than anticipated. It was both effective and inexpensive, superior in many ways to the citric acid it replaced.

In the years that followed, EDTA synthesis was further refined and modified both in Germany and in the United States. Marketed under many trade names, EDTA's commercial uses expanded as researchers perfected its ability to leech toxic heavy metals, such as lead, from biological and chemical systems.

Today EDTA and other chelators are widely used in homes as well as factories—in hundreds of everyday products. Few consumers realize that were it not for the chelating effects of household detergents, dirtied wash water would not drain free of scum. There would be an unsightly residue left around washbasins and tubs.

For all its industrial success, it was not until World War II that the potential therapeutic benefits of chelation were realized. Government concern with the possibility of poison gas warfare triggered a mammoth search for suitable antidotes.

The search ended when a team of English researchers, headed by Professor R. A. Peters at Oxford, implemented the chelation principle using BAL (British Anti-Lewisite, a chelating compound), which rendered arsenic from poison gases less harmful.

At the end of the war, chelation therapy was introduced into the medical arena. During the 1940s it became the routine treatment for arsenic and other metal poisonings. Poison gas fears proved to be unfounded, but a far graver public threat was developing: the real possibility of radioactive fallout that might contaminate mass populations should the atom bomb, then a hush-hush project, become a reality. Most radioactive contamination occurs in the form of isotopes of metallic ions, which can be chelated.

Americans, fortunately, were spared both eventualities, and the medical use of chelation was not intensively researched until the early 1950s. A group of workers suffering lead poisoning in a battery factory in Michigan were successfully detoxified. The chelating agent used was EDTA, found by American scientists to be more effective, with much less potential for adverse side effects than the British compound.

Next, the U.S. Navy adopted EDTA chelation therapy for sailors who were poisoned by absorbing lead while painting ships and other naval facilities. By the mid-1950s, it was becoming the accepted "treatment of choice" for lead poisoning in children and adults and, as of today, it still is.

Somehow a myth came about in years past to the effect that benefits of chelation therapy in heart disease were first observed in patients treated for lead toxicity. That is not true. EDTA chelation therapy was first used to treat heart disease by an eminent cardiologist and chief of research at the Providence Hospital in Detroit, Michigan.

Norman E. Clarke Sr., M.D., F.A.C.C., M.Sc., L.L.D., deserves credit as the originator of EDTA chelation therapy for treatment of atherosclerosis. He pioneered this treatment and, by acting as its principal spokesman, he kept the concept active and under study during the first two decades of its use. In the early 1950s, Dr. Clarke hypothesized that because EDTA binds calcium (and calcium is a substance deposited in arterial plaque) EDTA might reverse arterial blockage from atherosclerosis. Dr. Clarke and his associates were the first to perform clinical studies to test that hypothesis in patients with heart disease. It worked! But we now believe that chelation's removal of calcium may be relatively unimportant and that the binding of other metals that poison tissues and amplify oxygen free radicals may be more important. Almost half a century ago, Dr. Clarke came to a similar conclusion.

Dr. Norman E. Clarke Sr., with Robert Mosher, Ph.D., chemistry, conducted the first extensive studies with intravenous EDTA, determining safe dosage and manner of treatment. Dr. Clarke eventually came to believe that EDTA had its primary effect by acting on a variety of metals, allowing cells to resume normal function.

Dr. Clarke was an exalted scientific researcher and innovator. Among his many other contributions, he brought the first privately owned electrocardiograph machine (ECG) to Detroit. He authored many scientific papers, in addition to those on chelation therapy. He continued in active practice until the age of eighty-seven and lived to be ninety-two, clear of mind and with continued interest in chelation therapy.

Every study ever performed using EDTA chelation therapy to treat atherosclerosis has been positive. There are no negative studies, despite erroneous allegations to the contrary from critics of this therapy, as is made clear in chapter 10.

Patients treated with EDTA chelation therapy are able to walk farther, with less chest or leg pain than they had previously experienced. Those with angina are able to exert themselves without discomfort. They tire less easily and have much improved physical endurance.

Such dramatic benefits in patients with atherosclerosis are obviously related to increased blood flow through or around blocked arteries.

Intrigued by these new developments, cardiologists began to investigate and research the possibilities of chelation as a therapy for circulatory ailments throughout the body caused by atherosclerosis and related disorders. Early findings were encouraging and duly reported in the American medical literature beginning in the mid-1950s.

Two early researchers were Dr. Albert J. Boyle, professor of chemistry at Wayne State University in Detroit, and Dr. Gordon B. Myers, professor of medicine at the same university. They conducted research in association with one of the best-known cardiologists of his day, Dr. Norman E. Clarke Sr. Dr. Clarke was one of the earliest clinical researchers to prove benefit in patients with coronary heart disease.

Working at Providence Hospital in Detroit, they took on the "basket cases"—people so incapacitated by atherosclerotic cardiovascular disease that they were considered beyond help. And those patients improved. Following chelation, treated patients enjoyed a remarkable return of cardiac function and a reversal of disabling symptoms.

Clinical studies continued, and published reports consistently described clear signs of improved coronary circulation and heart function in most atherosclerotic patients after chelation. The findings—that patients had improved skin color, a return of normal temperature to cold extremities, improved muscular coordination and brain function, improved exercise tolerance without angina or shortness of breath, and a reduced need for nitroglycerine and pain relievers—were duly published in scientific journals. There have been dozens of such positive published clinical reports to date.

By 1964 the world's medical literature contained many scientific observations confirming those earlier findings. In that year, the distinguished Alfred Soffer, M.D., associate in medicine at Northwestern University Medical School and the former director of the Cardiopulmonary Laboratory of Rochester, New York, writing in his book, *Chelation Therapy*, stated that atherosclerotic patients suffering with leg pain from occlusive peripheral vascular disease appeared to benefit from repeated administration of EDTA, especially those patients with diabetes.

Clinical trials testing the effectiveness of EDTA chelation therapy in the treatment of arterial occlusion continue, and substantial progress has since been made. Two recent studies were performed by H. Richard Casdorph, M.D., Ph.D., assistant clinical professor of medicine at the University of California Medical School in Irvine, and by Drs. E. W. McDonagh, C. J. Rudolph, and E. Cheraskin.

Those and many other studies demonstrate clear and statistically significant increases in blood flow following treatment. Objective measurements were made before and after EDTA chelation therapy, using individual patients as their own controls.

Dr. Casdorph, utilizing sophisticated new noninvasive radioactive isotopes, demonstrated a statistically significant improvement of heart function and a highly significant increase in blood flow to the brain in patients with atherosclerosis. Precise measurements of cardiac ejection fraction (the percentage of blood pumped from the large chamber of the heart with each contraction) were determined before and after chelation therapy. Similar isotope techniques were used to confirm increased blood flow in carotid arteries and in the brain itself following chelation. The statistical probability that measured improvement could have been due to pure chance was less than one in ten thousand.

Dr. McDonagh and his colleagues in Kansas duplicated the Casdorph brain blood flow study results using a different technique. By varying pressure on an eyeball, it is possible to determine the pressure of arterial blood flow to the back of the eye. Since the artery supplying the eye is a branch of the internal carotid artery to the brain, blood pressure within the eye directly correlates with brain circulation.

Patients were used as their own controls, with measurements taken before and after chelation therapy. Improvements were also found by McDonagh to be highly significant.

These two excellent studies of blood flow to the brain were performed independently by different researchers in different locations, using different measurement techniques. The scientific community traditionally accepts the results of important new findings when results have been independently confirmed by separate researchers in unaffiliated facilities. These two independent studies followed scientific protocol, demonstrated an objective measurable effect of EDTA chelation therapy (increased blood flow to the brain), and served to substantiate what individual chelating physicians have observed independently.

Even without such sophisticated tests, we could presume that there was increased blood flow by sight and by touch. Patients routinely get their color back; their once-pasty complexions develop a healthy glow. Cold limbs regain warmth. Icy toes and icy fingers warm up. All this takes place in addition to a dramatic reduction of symptoms resulting from diminished blood flow.

Clinical results are consistently impressive. In the majority of cases, patients suffering the catastrophic effects of atherosclerosis (coronary artery disease, blockage of arteries to the brain causing stoke and senility, high blood pressure, peripheral vascular blockage of arteries to the legs, early gangrene, various types of arthritic and other related disorders) experience improved health. They regain lost physical and mental functions. They begin to "live" again.

As of this writing, more than a million patients have received in excess of twenty million chelation treatments in the United States alone, and not a single proven fatality caused by chelation has occurred when the treatment was properly administered and supervised.

Chelation therapy does not correct defective heart valves. Patients with valve problems often do, however, improve after chelation because of better heart function and increased coronary artery circulation. The heart works better, the pumping action is stronger, even if the valve remains unchanged. Patients feel better after chelation and, if valve surgery is necessary, the risks of a heart attack or stroke as a complication of surgery are decreased.

If the reader is mystified at this point, no wonder.

If chelation therapy is not old and discredited—and not new and untried—then what is going on?

If a safe, effective, tested, legal, nonsurgical treatment that seems able to reverse the symptoms of atherosclerosis and improve blood flow exists, then why haven't you heard of it?

2

THE MAKING OF A CHELATION DOCTOR

I didn't set out to be a chelation doctor—or any kind of medical specialist, for that matter. After completing training at Harvard Medical School, all I wanted was a traditional family practice. No fancy surgery, no Nobel Prize-winning research, not even a six-figure income and four cars in the garage.

Ambitions? Sure. I hoped to find a nice town where I could be a physician to whole families, treating all kinds of everyday ailments and complaints.

After some years as a naval flight surgeon, I spent six years as a family practitioner in Southern California in group practice. And then, while in Los Angeles attending a medical meeting in mid-1972, I heard a tale that was to change my approach to medical practice. It began with an innocent enough invitation.

"If you're not busy after dinner, drop into Room 1272."

"What's up?"

"George wants to tell a few of us about something new he's stumbled on—it's pretty revolutionary."

I was quick to accept, for the "George" referred to was an eminent and respected physician, a board-certified ear, nose, and throat specialist, and chief of otolaryngology at two hospitals. If George had something revolutionary to report, I wanted to hear it.

"In case you're wondering," our host began later that same evening, "what you've heard was correct. I did suffer with severe angina last year and was warned of the imminent danger of a serious heart attack. As a matter of fact, a few Long Beach doctors are surprised I'm still alive.

"As you can see, I'm back at work full time, and feeling a hell of a lot better than anyone, including myself, would have thought possible six months ago. The way that came about is what I want to tell you about."

Like all too many victims of advanced atherosclerosis and coronary artery disease, George first knew he was seriously ill when he was suddenly seized with severe chest pains.

As he told it: "There I was, enjoying a round of golf, when all at once, I felt as though an elephant had jumped on my chest."

The doctor became a patient. All the appropriate diagnostic procedures were performed at one of the nation's top medical school hospitals: electrocardiograms, treadmill tests, a coronary angiogram, the works. The results were not consoling.

Doctors usually level with one another.

"It doesn't look good, George," the cardiologists said, reporting the angiogram had revealed plaque blockages obstructing the left main coronary artery and other arteries as well. Their recommendation was a triple coronary artery bypass.

"No time to waste," the specialists agreed, pointing out there was impending danger of total occlusion and a real possibility of sudden death.

"Ouch," George said.

Physicians are as apprehensive as laymen when it comes to going "under the knife"—perhaps more so, being all too familiar with what can go wrong with major surgery.

Coronary artery bypass is a particularly sobering prospect, inasmuch as it entails any number of hazards. During the 1970s, a significant percentage of patients—as high as 10 to 15 percent—died as a direct result of the surgery. Survivors had no assurance of long-term improvement. No one could say for sure that the bypassed arteries would not degenerate further, or that blockages might not develop at the site of the graft. It was a high-risk surgical procedure with uncertain benefits. Today the procedures have been improved and a lot fewer people die on the operating table, but the benefits are still not certain.

For George, however, there did not seem to be any choice. Then fate intervened.

The Christmas holidays were looming, and the Red Cross was unable to supply the seven standby units of blood required until after New Year's. George chose not to remain idle during his two-week reprieve.

"I talked about my condition to everybody I could get hold of—doctors, lawyers, my golfing buddies, my accountant."

A week later, a physician George described as "a very dear colleague who went to medical school with me, interned with me, and later became a prominent New York internist," called to ask, "Ever heard of chelation?"

"How do you spell it?" George asked, and then headed for the medical library.

"I'm sure you fellows are going to be as amazed as I was," he told us. "There are a dozen or so pretty impressive clinical reports on this chelation therapy that I'll bet most of you have never heard of."

Further investigation by George led to his decision to postpone surgery long enough to give this apparently safe, painless, noninvasive therapy a chance to work. Intent on being treated by a doctor experienced in chelation, George traveled more than two thousand miles to one of the few clinics in the country then specializing in the treatment.

"You're going to find the rest of this hard to believe," he warned.

"After only ten of those treatments, my angina disappeared—completely! Imagine how I felt. Before I started treatment, I couldn't walk a dozen steps without severe angina. But, I swear, I haven't had a chest pain since.

"What's more, I saw dozens of other patients, even worse off than myself, healed from conditions you and I think are incurable. One arrived with a gangrenous leg, and I saw that man's leg heal and begin a return to normal.

"I wondered whether I could believe my own eyes: people checked in with chronic diabetic ulcerative lesions and gangrenous lesions that showed visible signs of healing in ten to twelve days."

Ordinarily somewhat reserved, George waxed on with uncharacteristic enthusiasm, crediting chelation with astonishing cures, not the least of which was his own.

"I now carry a full work load. I do approximately ten to fifteen surgeries every week, and these are microsurgeries of the ear. I carry a full practice. I play golf. I swim twenty laps in my pool every day, and I cannot speak with any but the greatest praise for the men who are attempting to make chelation an accepted form of therapy."

If it were anyone but George reporting such wonders, I doubt that any of the physicians listening, myself included, would have given much credence to so fanciful a tale. But it was impossible to doubt his sincerity, integrity, and even more important, his medical credibility.

We had dozens of questions:

"What did the treatment involve?"

"A series of intravenous injections."

"How many treatments—how long?"

George had taken a series of twenty treatments, each lasting three to four hours, at the clinic, and had since learned to self-administer. He continued the intravenous injections with the help of a nurse in his own office.

"Was it legal?"

"Yes."

"Was it painful?"

"No."

"Was it safe?"

"Yes."

"Did it work in every case?"

"What does?"

"Why haven't we heard about it before?"

The key question.

George did not have a good answer.

Chelation therapy has been relegated to a medical no-man's-land ever since the chelating agent—EDTA—which had formerly been accepted by the Food and Drug Administration (FDA) for the treatment of occlusive vascular diseases (such as angina) lost this status because of an unfortunate, uncontested change in federal regulations.

Previously it had only been necessary to prove EDTA's safety. That was no problem. To meet the newly regulated FDA requirement for proof of effectiveness for the treatment of the atherosclerosis would require that tens of millions of dollars be spent on research. No drug company was willing to underwrite such an expense because the patent on the substance had already expired. And there was no longer hope for recovery of costs.

Chelation with EDTA was—and still is—an important FDA-accepted treatment for lead poisoning and dangerously high blood calcium, but physicians electing to use the treatment for any other purpose do so at the risk of flouting FDA guidelines and the recommendation of the American Medical Association. While a physician may legally use any FDA-approved drug however he deems suitable, he does risk increased vulnerability to malpractice suits (and the frowns of his colleagues) when he defies established recommendations.

George was optimistic that EDTA chelation therapy for atherosclerosis could not remain in limbo for long.

"It's too good to be ignored," he said. "All that's needed is for a reputable group of physicians to get moving on this—to study up, start treating patients, and document its value."

George made it clear that he hoped there were doctors among the group invited to witness his testimony who would "get the chelation ball rolling." It was an irresistible challenge. Even if George were not so well respected, we would have been eager to find out whether chelation therapy would hold up to more intensive scrutiny.

It was a lucky time for anyone interested in a novel medical modality to be practicing in California, where so many new ideas—political, social, and scientific—seem to take root and grow. The informal study group of about a dozen doctors that formed that night (and eventually spawned the American Academy of Medical Preventics, which later became the American College for Advancement in Medicine) began meeting regularly. There were seminars, lectures, and physician training sessions. Doctors learned to improve chelation techniques by apprenticing to each other.

The more I learned, the more intrigued I became. I read dozens of monographs and hundreds of articles in various medical journals published here and abroad. I talked at length to pioneer chelation practitioners and interviewed many of their patients. I soon learned to administer the therapy. I was my own first patient, taking chelation therapy myself. No problems at all. I shared experiences with colleagues doing the same, and all of us agreed that chelation therapy was so worthwhile that it would soon be a universally accepted and practiced treatment.

That was in 1972. It hasn't happened yet. As it turns out, George was an astute scientist, but a rotten prophet.

I chose to leave the scene for a brief time, when my long-standing involvement with public health and preventive medicine led to a position with the U.S. Public Health Service. I became chief-of-staff at the Talihina, Oklahoma, Indian Hospital, where I supervised the medical care of twelve thousand Choctaw Native Americans. Although that assignment removed me from active involvement with chelation, my interest never dimmed. During that stint, I kept in touch with chelation practitioners, stayed current with the new literature, and took annual refresher courses. Finally, in 1976, it seemed the right time to settle down somewhere with my wife and our four young children.

Through the medical grapevine I heard of an opportunity to set up a full-time family practice in a small town, Trout Dale, in southwestern Virginia's Blue Ridge Mountains. The Mount Rogers Clinic (a modern out-patient facility) was to be my headquarters, and I would be the only doctor within a twenty-mile radius.

Trout Dale (population 250 but with 4,000 people residing within ten miles of my office) isn't really "near" anywhere. It is more than an hour and a half drive to the closest big city, and the nearest movie and supermarket are thirty-five minutes away by car. I was not just a family doctor now; I was a country doctor.

Some days were frantic; others were not. Sometimes I worked round the clock; other times I could have gone fishing and no one would have noticed. I was on call twenty-four hours a day, seven days a week. The Mount Rogers Clinic was not only my office but my home as well, providing living quarters for my family. Oh yes, we were also the local pharmacy.

It was the perfect setting for my one-on-one, prevention-minded brand of medicine. I'm a holistically oriented doctor—a maverick in the eyes of some of my colleagues. I use

15

whatever treatments work best for my patients, even if not yet endorsed by the more conservative medical groups. For example, I have long advised dietary changes and vitamin supplementation to enhance other remedies. Long before it was popular, I was encouraging patients to exercise regularly and to upgrade their nutrition.

At first, my medical practice focused mainly on the routine: ear infections, broken arms, pneumonia, and gout. In my early career I did major surgery and delivered more than six hundred babies. There was still the occasional car wreck, accidental poisoning, serious burn, personal assault injury, on-the-job accident, or other medical trauma. I made lots of house calls. It made for a full life—a typical country doctor existence—until I introduced chelation therapy.

Rural Virginians develop vascular diseases just like big city folk, and they are no more eager than their urban counterparts to undergo surgery. One of the first patients I chelated was a local businessman whose heart specialist had told him, "Get your affairs in order. You won't last long."

When after a full course of treatments, he was back at work, putting in an eight-hour day, acting twenty years younger, we lost our seclusion forever. On the basis of initial successes, patients sent others, many far from Trout Dale. Our daily caseload soon included men and women from up and down the Eastern Seaboard. Like born-again evangelists, our recovered chelation patients regarded it as their mission to spread the word.

And for good reason. Many considered their cure nothing short of a miracle. Detailed case histories time and again document remarkable recoveries from an almost endless variety of so-called hopeless degenerative ailments.

In 1996 I founded a second clinic, the Mount Rainier Clinic in Yelm, Washington, near Olympia and Tacoma. That's where I now live and practice.

Chelation therapy is not a cure-all or an "elixir of life," but it can do things that no other therapy or surgical procedure presently available has any hope of achieving. The next chapter is an illustration of that fact.

3

THE STORY OF JUDY

She came to my clinic in rural Virginia because she had nowhere else to go, and she was in dreadful—almost terminal—shape.

Like many of my patients, Judy had rejected the advice of her physician to undergo immediate surgery. She had developed gangrene in her right foot and was advised to have it amputated. Judy was not quite fifty years old, a working woman, and a grandmother. She had suffered with increasingly poor circulation caused by atherosclerosis for a good many years. She had lived by her wits all her life, and she was determined not to lose one of her limbs—even if she had to die to keep it. So she came to me.

According to most cardiovascular surgeons and specialists, there is only one remedy for a limb becoming blackened with gangrene resulting from end-stage atherosclerosis: cut it off before it poisons the whole body.

I do not accept that in every case. Especially when the gangrene has just begun, chelation therapy can often restore blood circulation and promote healing.

Amputation, on the other hand, neither treats nor eliminates the root cause of ischemic gangrene (blocked circulation and oxygen starvation), which will most likely recur at a future time higher up in the severed limb or in the other leg.

Rather than remove the gangrenous leg, I prefer to try chelation, not just because this is a less mutilating approach to a devastating ailment, but also because it allows more options. If successful, the limb will be restored to normal, or at least acceptable functioning. Minor surgery may be needed to trim away only the dead tissue. If chelation should prove unsuccessful, amputation is still available as a last resort.

In Judy's case, ischemic gangrene had surfaced in her right foot, which had begun ulcerating three months before. Soon the foot was mottled with black patches. Three toes, then the heel, developed festering wounds that oozed a dark, smelly discharge. For Judy, pain throbbed unceasingly, making it impossible for her to work, walk, even to sleep.

Attending doctors offered Judy a Hobson's choice: your foot or your life. But Judy would have none of it. She suspected the first excision would most likely lead to another. And another. Her atherosclerotic wounds evidenced a progressive disease. She had heard this referred to as the salami technique, "a slice at a time." She told me, "I'm not going to have any more surgery. If the pain gets too bad, I'll blow my brains out."

Judy's decision resulted from a long and expensive medical history that had left her physically and emotionally scarred. After filing for bankruptcy, because of huge medical bills, she had had her fill of hospitals, operations, and doctors, however well intentioned their advice. For a solid year, Judy's life had entailed a grueling series of expensive, unproductive medical consultations and confinements. Despite everything, her health had continued to deteriorate.

She could hardly recall a time when she could get around pain-free. For several years, whenever she climbed stairs, took a walk, or stood in line, she would experience leg cramps and throbbing pain. "When I complained to doctors, they would tell me, 'Don't worry. It's all in your mind.'" The pain in Judy's legs continually grew worse.

"Then came the day my body finally gave out. I just collapsed." When Judy regained consciousness, she was in intensive care at DePaul Medical Center in Norfolk, Virginia. Her left side was numb and her doctors believed she had developed embolisms, small blood clots or tiny pieces of atherosclerotic plaque that break off from an arterial wall and circulate through the bloodstream until they become lodged in the narrow, downstream segments of blood vessels. Three angiograms were performed. "A ghastly experience," Judy recalls.

The diagnosis was grave: total occlusion of the left subclavian artery with far-advanced blockage of the arterial systems to her left arm and to her right leg. Translated, Judy had end-stage hardening of the arteries—critical blood vessel blockages that severely limited circulation throughout her body and especially to her extremities.

Then, as now, the commonly accepted surgical treatment for Judy's ailment was arterial surgery. There were three operations in Judy's case: the first, to replace the descending aorta and iliac arteries with a man-made Dacron "Y" graft; the second to bypass the blockage in her right leg; and the last for the left arm. Assured that there was no alternative, Judy reluctantly agreed to the first operation.

"It almost killed me," she reports. "I went into cardiac arrest twice."

After that ordeal, Judy was too weak for further surgery. She was discharged from the Norfolk hospital, prognosis "improved" but still suffering from occlusive peripheral vascular disease, a threatening condition that inevitably deteriorates. In Judy's case, it wasn't long before she knew she was destined for more trouble. Her pains were as bad as ever. The surgery had brought no relief at all. The branch of the aortic "Y" graft leading to her right leg had blocked off once again.

"The slightest activity was a pain-filled, huff-and-puff effort," Judy remembers. "I felt I was living on borrowed time. I stopped making plans more than one week ahead."

She maintained her regular medical checkups. "Each time I questioned my doctors about what to do next, the answer was always the same: 'More surgery.'"

At this time, Judy was truly desperate for alleviation of her disabling pain, any kind of alternative to the dreaded arterial surgery. As it happened, our paths crossed. I was the featured speaker at a health seminar in Roanoke. Judy was there and she buttonholed me in the parking lot.

I could see right away how distraught she was, how frantically she wanted an empathetic listener. Judy rattled off her medical history and expressed interest in my talk about the healthful benefits of diet, exercise, and vitamin supplements. But what had really caught her attention had been my brief remarks about chelation therapy and its effectiveness in reversing the symptoms of atherosclerosis.

"I've never heard of chelation," she said. She wanted to find out more. I mailed her an information packet. She wrote back saying she did not think she could afford any more medical care.

"Your clinic in Trout Dale," she explained, "might just as well be in another country as two hours away. I'm broke, depressed, and have run out of health insurance. Constant illness has forced me to change jobs a lot and lose benefits. And I still owe for old medical bills. It hardly seems the time to take leave from my job again."

But soon thereafter, Judy was forced to rearrange her priorities once more and put survival at the top of the list.

"I began to suspect it was the beginning of the end," she remembered. What triggered her presentiment was a sore big toe, a very sore big toe, which refused to get better. "I tried all the usual treatments, but it just got blacker and wouldn't heal.

"Then, the toe began to seep. And ooze. I had been swimming a lot, but I gave it up because I was afraid that whatever I had might be contagious. Then the rest of the foot began to ache and look funny." It was time for Judy to see a doctor again.

Her worst fears were confirmed. The arterial graft in her right leg had completely blocked. Almost all circulation to the foot was gone and ischemic gangrene had set in. The consulting surgeon did not mince words; amputate right now, just below the knee, and stay alive.

"More surgery? Never. No way," Judy replied.

The surgeon shrugged, as much as to say, "It's your funeral."

And Judy was resigned to her fate. Her condition worsened with each passing day. She could no longer walk; she had to be carried upstairs where she worked, then hobble painfully to her desk. "It all seemed so hopeless," she recalls. "I was fully prepared to die. As a matter of fact, it seemed a welcome solution."

It was about this time, that I chanced to call Judy's pastor in Roanoke. Judy had been on my mind ever since our conversation in the parking lot and I heard of her wretched status. Then I phoned to invite Judy to Mount Rogers.

"Judy, I can't make any promises, but I think we may be able to help. How soon can you get here?"

"Do you have any idea how sick I am?" she asked. "Don't you know I haven't got any money?"

"Don't worry about that now," I said. "Just come."

And so Judy did come and started chelation therapy in April. Eighteen months later she took leave of us in Trout Dale, a happy and much healthier woman and the most miraculously recovered patient who ever came under my care.

What did I do for Judy? What did I do to her? First of all, I accepted the risk of trying chelation therapy on Judy, who was nearly moribund during the first six weeks at our clinic. But many chelation patients come to us in close-to-death condition: "You're my last hope," we hear over and over again. In those cases, where all other treatments have failed, the patients and I often share a "nothing to lose" attitude.

No doctor likes to initiate treatment after the patient is so seriously ill that he or she hovers at death's door. But we try.

Judy came to us at the last moment because, like so many others, she had never heard of the chelation alternative until it was (almost) too late.

For weeks she showed no signs of improvement with chelation. Finally, the treatments began to take hold. Her progress was slow but steady thereafter and always inspirational.

"No one here is as sick as I was," she would greet other arriving patients, "and look at me now. I'm alive."

To this day, Judy is a local legend around the Mount Rogers Clinic. She remains our most striking example of what can be achieved with chelation once blood circulation is restored in blocked arteries.

A prominent cardiovascular surgeon on the teaching faculty of a nearby medical school examined Judy in Winston-Salem, North Carolina, when she was about to leave us.

"This lady has a remarkable story," wrote the doctor. "She has completely healed necrotic places on her feet, which were obviously caused by her atherosclerosis."

This physician called me to say, "There is nothing I could have done any better, even with vascular surgery. Most patients with her severity of disease would have long since had to have their legs amputated."

The dramatic case history and recovery of Judy, impressive as it is, is but one of tens of thousands of similar striking reports in the medical files of chelating physicians, which detail equally striking recoveries.

Judy continues to do well ten years later. She has remarried and leads an active and exciting life.

With each passing year over the last three decades, increasing numbers of trail-blazing physicians (more than one thousand in the United States alone) have begun practicing chelation. The results? No one has done a statistical survey, but I do not personally know of any physician who has used EDTA chelation therapy for any length of time, in the actual treatment of patients with atherosclerotic vascular disease, who is not convinced of both its effectiveness and safety. Most take the therapy themselves, if only as a preventive.

But, to the continued frustration of doctors like myself, who year after year demonstrate the clinical effectiveness and safety of chelation therapy, this treatment is still ignored—derided, even—by most mainstream physicians. While there is nothing about chelation that smacks of quackery (it cannot be administered by anyone other than a fully licensed physician able to use prescription medicines), it is still shunned by the medical establishment. Why?

An inherent conservatism within the medical profession can serve to protect the public from ineffective, fraudulent, and potentially harmful therapies. On the other hand, it also slows the acceptance of new medical advances, sometimes for decades.

In this scientific age, a remedy that makes people "feel better" stands little chance of being accepted as valuable unless large-scale, prospective double-blind crossover studies confirm what the clinician has observed firsthand. Medical scientists are conditioned to consider any other type of evidence as unreliable. Testimonials from practicing physicians are treated with disdain. The majority of the medical community, in its smug self-satisfaction, has forgotten that for the most part the laboratory in medicine has followed the kitchen. Many a medical breakthrough was first reported anecdotally. For example, country doctors were feeding their "rundown" patients liver soup long before scientists could identify the cause or cure of pernicious anemia.

In addition, chelation therapy suffers from the "N.I.H." (not invented here) condemnation. Medical advances are supposed to emerge from the richly funded clinical laboratories of the nation's leading medical centers or from grant-rich government supported research projects at academic ivory towers—not from the everyday experiences of private physicians.

I believe there is still another and even more important reason chelation therapy is still struggling for recognition. Until very recently, even its staunchest advocates were unable to offer a scientifically defensible explanation for the observed benefits. Even now, we are still not certain about how this therapy produces its benefits in the body.

While there have been thousands of articles published in scientific journals about chelation therapy and the usefulness of EDTA for a variety of conditions, most of the articles concerning treatment of atherosclerosis and age-related diseases have concentrated on documenting how well the therapy works, but have given short shrift to the question of how it works. For more than four decades, the dynamics of the treatment have remained a mystery because most reports had focused on its usefulness for lead toxicity. Only in the 1980s had science progressed to the point where the complex biochemistry could begin to be unraveled. I explain this more fully in future chapters.

Breakthroughs in an area of medicine known as free radical pathology now enable us to provide a coherent, scientific rationale for at least some of the many and diverse benefits chelation practitioners have long claimed to have observed.

The free radical concept may be as profound in its implications as the germ theory developed a century ago. Just as the germ theory provided a scientific basis for effective treatment of the major killers of the nineteenth century—the infectious diseases—so the free radical concept provides the groundwork for treatment and prevention of the major killer diseases of this century: atherosclerosis, heart attack, stroke, senility, arthritis, and cancer. We are now beginning to understand why so many chelated patients report feeling as though their treatments turned back their aging-clocks and restored much of their youthful vigor. For the truth about chelation therapy, as we now understand it, is this: It seems to partially reverse and then slow the very aging process itself.

4

CHELATION THERAPY: WHAT IT IS, WHAT IT DOES, HOW IT WORKS

Only after the advent of an "acceptable" scientific explanation does an innovative, therapeutic approach stand a chance of being acknowledged by the medical community. Until a physician can explain why a patient has improved with a chosen treatment, his observed cures often have little credence among scientists.

A case in point follows. In 1850 Dr. Ignaz Semmelweiss, a Hungarian obstetric physician, reduced maternal mortality rates following childbirth in his Vienna hospital from 25 percent to less than 1 percent simply by requiring attending doctors and medical students to wash their hands. Until his death in 1865, Dr. Semmelweiss was plagued by the failure of his colleagues to accept the lifesaving benefits of his procedure. It was not until the very year he was buried that an English surgeon, Dr. Joseph Lister, introduced the principle of antisepsis to surgical procedures, drawing upon Semmelweiss's observations and the work of Louis Pasteur, who proved that microscopic bacteria exist all around us and even permeate ordinary air. By the time of Lister's death in 1912, hand washing and antiseptic surgery were commonly accepted practices. But it had taken fifty years, in no small part because doctors were unwilling to concede that they themselves had been transmitting infections on their unwashed hands.

The perfectly valid recommendations of these two physicians might have been ignored even longer if Pasteur's explanation—the germ theory—had not been accepted at long last by the medical community. Now even school children know about germs and understand that cuts, scrapes, and bruises must be kept clean.

In the one hundred years since Pasteur made medical history, there has not been another discovery so profound until the recent concept of free radical pathology as an important underlying factor in degenerative diseases of aging.

The free radical concept of aging, first postulated by Dr. Denham Harman in 1962, is one of this century's major medical advances, inasmuch as it provides a cohesive explanation of seemingly contradictory epidemiologic and clinical evidence concerning age-related pathology.

Decades of controversy have pitted scientific theories against each other. One has blamed cholesterol and diet, another the environment, another homocysteines; still others point to stressful lifestyles to explain the ever-increasing numbers of Americans being felled by heart attacks, strokes, and cancer. Some have said it's because we are living too long. Still others claim it is because we are living too well—not enough hard work and exercise. Most blame some combination of all the above, plus some as yet little understood factors such as genetic tendencies or overexposure to environmental chemicals or radiation.

It was not until comparatively recently that researchers had access to sophisticated technology enabling them to trace biochemical pathways at the most elemental molecular level to unravel the diet-stress-metabolic-environment-radiation puzzle.

The dramatic result has been a breakthrough in our understanding of the causes of the major diseases of aging, which has provided a sound scientific basis for the use of nutrition and modification of stress and harmful lifestyle factors in the treatment and prevention of degenerative ailments. It may even be possible to slow the very aging process itself. The free radical theory of aging introduces a fresh principle: whether or when one succumbs to a degenerative disease (such as atherosclerosis) depends largely on the body's ability to defend itself against ongoing free radical attack. While the maximum life span for a species or an individual appears to be predetermined by genetics (DNA coding), free radicals have a decided influence on whether that limit is reached or not.

We can now postulate a sound scientific basis for EDTA chelation therapy. But before we get into that, let's cover some basics—what chelation is and what it does—before we discuss how we think it works.

Here are the questions people most often ask when they first learn of chelation, and the answers I give them:

How do you pronounce chelation?
With a hard "k" sound: key-LAY-shun.

What is chelation therapy?
To repeat what was said in the introduction, it's a medical treatment that improves metabolic function and blood flow by removing toxic metals (such as lead and cadmium) and abnormally located nutritional metallic ions (such as iron) from the body. It can also redistribute essential elements to more functional sites in the body, even without removal. This is accomplished by administering a synthetic amino acid, ethylenediaminetetraacetic acid (EDTA), by an intravenous infusion using a tiny gauge needle or flexible Teflon catheter.

29

The verb *chelate* is derived from the Greek noun *chele*, which is the claw of a crab or lobster. Thus chelation, a natural process, is the pincerlike binding of chelating substances to metallic elements.

Why is chelation a "natural" process?

Chelation is a basic life process that enables all growing things and living organisms (including plants) to assimilate and make use of essential inorganic metallic elements. Chelation is nature's marriage ceremony: It weds two substances from totally different chemical worlds—the organic and inorganic—into a compatible working partnership.

Chlorophyll, the plant-greening pigment, is a chelate of magnesium. Hemoglobin, the oxygen-carrying pigment of red blood cells, is a chelate of iron. The chelation process is involved in the formation and function of many enzymes—those protein catalysts that control most of your body's vital biochemical functions.

Industry has made good use of the chelation principle. Chelates are used in household cleaners, for example, to reduce the "ring around the bathtub" scum buildup. Chelates binding with magnesium and calcium soften hard water. In much the same way, EDTA removes and rebalances abnormal metal ions in the body, reducing the production of free radicals (which will be explained later in this chapter) and improving enzyme function. This can be looked on as preventing conditions under which another type of scum builds up within the walls of blood vessels. This is an oversimplified explanation of what happens, but will help initiate an understanding of the process.

How does the EDTA work?

EDTA is a small, protein-related, amino acid molecule with unique and valuable therapeutic properties—one of which is its powerful attraction for loosely bound metals, including those that speed free radical damage.

Toxic metals, such as lead and cadmium, are double trouble. They disrupt basic metabolic processes, inhibiting vital enzyme function, thus increasing free radical damage. EDTA binds tightly to many other abnormally placed and potentially toxic metal ions. It forms a chemical bond that facilitates removal of the metal from sites where it might interfere with normal metabolism or otherwise enter into an undesirable chemical reaction.

Isn't it dangerous to have essential nutritional metals such as iron, copper, and zinc removed from the body?

When they are where they are supposed to be, essential metals are so tightly bound to the sites of normal activity that they are not easily removed. The metallic ions most apt to be removed by EDTA chelation are those that are loose and free-floating. Because this does not hold true all of the time, we are careful to provide nutritional supplements to replace lost essential metals.

I've heard chelation described as a "chemical roto-rooter." Is that correct?

It's graphic and colorful, but it's also scientifically inaccurate. Hoping to make a difficult subject more understandable, a chelation doctor once described chelation as a kind of liquid plumber—a "chemical roto-rooter" that increases blood flow by reaming calcified plaque out of the arteries.

The explanation stuck, and it has haunted chelation specialists ever since. The "chemical plumber" theory is totally invalid, for although EDTA is certainly a calcium chelator, its affinity for calcium is far less than for other metals, such as iron, copper, lead, cadmium, manganese, etc. EDTA will quickly release calcium to pick up one of those other metals.

Furthermore, calcium, as we will see, plays but a secondary role in plaque formation. The amount of calcium removed by a chelation treatment is minuscule compared to the total amount in the body.

 What does chelation do for the patient?

Chelation therapy has been proven to increase blood flow throughout the body. It's been reported to improve liver function, improve blood cholesterol ratios, lower blood fats, reduce blood pressure, reduce leg cramps, improve vision, relieve angina pains, relieve symptoms of senility, heal ulcers caused by poor circulation, forestall heart attacks and strokes, relieve symptoms of arthritis, relieve symptoms of Parkinson's disease, improve memory, and reduce the incidence of cancer.

How much does blood flow increase as a result of chelation?

Enough to relieve symptoms in most cases, promote healing, and improve the quality of life.

It takes only very slight vascular changes to significantly alter blood flow. Of the thousands of miles of arteries in the body, most are so slender that they have an opening no wider around than a human hair. Most capillaries are so narrow that single blood cells must fold themselves in half to squeeze through. Such diminutive pathways block easily—and small changes may unblock them.

A long-established scientific law states that with perfectly laminar flow (fluids moving easily through a smooth-walled conduit) a mere 19 percent increase in the diameter of a vessel will double the flow rate.

But the arteries in most of our patients are not smooth— they are usually filled with irregular plaque and, as such, are turbulent conduits. Poiseuille's Law of Hemodynamics tells us that in the presence of turbulence, it takes something less than a 10 percent increase in diameter for a doubling of blood flow.

Those percentages are counterintuitive but can be proven mathematically and experimentally.

How does chelation increase blood flow?

EDTA is similar to aspirin insofar as we simply do not know all the mechanisms involved. Aside from its "buddy-up" action with toxic heavy metals and other abnormally situated metallic ions, at present we are uncertain of all its beneficial properties.

One published article lists more than twenty physiological and biochemical actions of EDTA within the human body, any one of which has the potential to dramatically improve physiological function and increase blood flow.

EDTA chelation therapy can, for example, improve arterial elasticity by reducing the number of cross-linkages in the connective and elastic tissues comprising the arterial walls. The artery walls would then be better able to relax and dilate. Blood flow could increase without otherwise altering existing plaque. Some such cross-linkages are caused by metallic ions bridging large protein and connective tissue molecules. Chelation therapy can also reduce nonmetallic connections that occur abnormally between large molecules, such as sulfhydryl (sulfur to sulfur) cross-linkages.

There is good evidence that EDTA exerts a beneficial effect on the metabolism of individual cells by enhancing the use of oxygen and other nutrients, even in the presence of compromised blood flow and diminished tissue oxygen. This effect has been documented experimentally.

EDTA restores the normal production of prostacyclin (the "Teflon" of the arteries), which prevents spasms, blocks clots, reduces platelet "stickiness," and improves blood flow, even in diseased arteries. Prostacyclin production turns off in the vicinity of free radical activity.

Which takes us to the most important effect of all: EDTA can reduce the localized rate of free radical reactions in blood vessels and elsewhere by removing accumulations of metallic catalysts of lipid peroxidation. And this benefit persists for a long time after chelation is administered.

Peroxidized fats (or lipids, as fats are called scientifically) are rancid—they have combined with oxygen via a free radical catalyzed reaction, releasing mutagenic substances that in their turn oxidize further, creating still more free radicals, and triggering a potentially out-of-control chain reaction.

To grasp the significance of the free radical concept, especially as it pertains to chelation with EDTA, some background understanding of basic biology is helpful. Your body is composed of about 60 trillion cells, each enclosed within an encircling cell wall or membrane.

Cells vary greatly in structure, as well as in function. Those in certain bodily structures—such as the skin, the linings of the intestinal tract, and the blood cells—are continuously being worn out and have the ability to replace themselves frequently. Other tissues, such as the brain, nerves, and muscles (including the heart), are largely made up of nondividing cells that, once worn out, cannot renew themselves. Over time, nonrenewable cells become increasingly damaged in the course of their activities: they age, they die, and they clog tissues and organs and biochemical pathways as cellular "rubbish." Although replaced as needed, renewable cells tend to degrade with each successive cell division and renewal.

When your cells are damaged, you are damaged. When your cells perform inefficiently, you perform inefficiently. When enough cells die, you die. How do cells get sick and die? One way is to succumb to free radical attack. Another is to have metal ions accumulate to toxic levels.

With each subsequent cell division, replacement slows and the healing process is retarded. This is called "cell senescence." As they're replaced, cells become increasingly weaker and defective. Skin cells and cells in hair follicles are replaced very frequently throughout life. Resulting deterioration is plainly visible to the naked eye, such that a close estimate of chronological age can be made at a glance. Wrinkling, sagging, and thinning of skin with age, and graying of hair are inevitable and clearly visible. Cell types that divide frequently throughout life also deteriorate at predictable rates. Essential trace metals are needed at every point in the cell replacement process, and they become abnormally distributed with age.

In heart muscle, for example, coronary artery blockage can cause nutritional trace elements to rise precipitously. In coronary artery disease, cobalt has been observed to increase 500 percent; chromium increased 520 percent; iron increased 400 percent; and zinc increased 280 percent. Although those are all essential, nutritional trace elements, they have a narrow margin of safety between useful and toxic levels. Three- to fourfold intracellular increases could poison metabolism. Realignment and redistribution of such essential trace elements, with augmentation of vital metalloenzymes, may be as important, or more so, than elimination of metals from the body.

A free radical is an oxygen molecule with an odd number of electrons in the outer orbital ring of one of its atoms, sometimes referred to as "reactive oxygen species"—or simply "oxygen radicals." Molecules (and atoms) normally contain an even number of paired electrons. Free radicals differ from normal molecules, ions, and molecular complexes by having an unpaired electron in their structure, a distinction that may seem trivial but that has enormous significance.

If one of the electrons in a pair becomes separated, an imbalance is created. That imbalance makes the resulting molecule (or atom) promiscuously unstable, violently reactive, and potentially very destructive, ready to aggressively attack any nearby substance, setting off further free radical reactions with explosive cell-destroying power. What free radicals actually do is combine with and react chemically with other molecules that were never meant to be interfered with. Just as outside of us oxygen produces rust on metal surfaces, so, inside of us, unbalanced oxygen molecules "rust" the body. The rust on the bottom of your car or on an old metal fence is nothing more than the oxidation of iron. Free radicals are continuously generated as a result of many essential chemical reactions that occur naturally in the body. Those reactions are necessary for life as part of the normal metabolic process. They're a byproduct of the normal use of oxygen, needed to burn fuel (food) and generate energy, and are produced in large numbers in the mitochondria, the cell's complex "oxygen reactor" or power plant. Free radicals are also generated in the endoplasmic reticulum (detoxification compartment) of liver cells, in white blood cells, and in other locations. The uptake and transport of oxygen to tissues by hemoglobin in red blood cells involves a free radical reaction.

At the same time that oxygen and fuel from food are processed in the mitochondria to generate the cell's energy requirements, a flux of free radicals is released. These raiders head for the nearest target—the fatty acids within the mitochondrial membranes—converting them into dangerous peroxidized fats, which then produce even more free radicals.

Uncontrolled, these free radicals can wreak havoc. They are deadly marauders that damage cells by breaking down delicate cell walls, by damaging important protein enzymes, by altering sensitive DNA chromosomes and genes in the nucleus, and by ravaging sensitive internal structures of mitochondria so they are no longer efficient energy producers. Free radicals have a life span of microseconds and their concentration in any location at any given instant is minuscule. But they'll attack anything in their vicinity with amazing speed.

To appreciate how deadly free radicals may be, consider this: Should you be exposed to excessive x-rays or gamma ray radiation, you'd be damaged by free radicals produced by that high-energy radiation. Any time your body is exposed to nuclear radiation—or excessive x-rays—it's free radical damage you must fear.

There is absolutely no way to escape ongoing exposure. Even before the nuclear age, man was constantly subjected to radiation from cosmic rays. One quart of ordinary air on a sunny day contains about one billion free radicals of a potentially dangerous form of oxygen called ozone. Radiation from the sun and stars continually filters through the atmosphere, subjecting our bodies to free radical exposure.

Contrary to what most people believe, it is not this external radiation that is most worrisome. Antinuclear pickets at the Three Mile Island nuclear reactor site might have been surprised to learn that even those persons living closest to the plant were damaged more by their own internally produced free radicals than by those caused by escaping radiation. We are all subjected to continuous internal radiationlike effects of highly reactive free radical molecules that are regularly produced within our cells in the normal course of daily life; this is a consequence of merely eating and breathing. How then, does man survive?

As you might suspect, nature has provided us with exquisitely designed survival equipment. Every cell that deals with oxygen is equipped with an antioxidant defense system that can quickly and efficiently scavenge and neutralize most free radicals generated normally in the body.

In a healthy body, free radical reactions are controlled but still allowed to proceed in an orderly fashion as needed for normal energy production and for detoxification of chemicals, germs, and foreign substances. Several enzymes (including catalase, superoxide dismutase, and glutathione peroxidase, in cooperation with other antioxidants) keep free radicals from running wild. When functioning properly, these enzymes, in concert with an elaborate system of natural free radical scavengers (including the antioxidant vitamins, C and E), dampen free radical chemical reactions, thus allowing the desired biological effect without unwanted cellular or molecular damage.

In fact, virtually the whole spectrum of known vitamins, minerals, trace elements, and antioxidants work together in harmony to form an elaborate antioxidant system. When an antioxidant is inactivated by a free radical, it's quickly regenerated by the next one in the cascade. That system is only as effective as the weakest link. That's why supplementation with a broad-spectrum product, containing thirty or more ingredients, is much more effective than taking just one or a few antioxidants, such as vitamins C or E by themselves.

Without effective antioxidant controls, there would be unrestrained free radical proliferation. Much like a nuclear chain reaction, these would cause out-of-control free radicals to be generated at an ever-increasing rate. When that happens, they disrupt cell membranes, damage essential enzymatic proteins, interfere with transport across cell membranes, and cause mutagenic damage to the genes and chromosomes. All such activities produce very sick—or even malignant—cells.

The damage is cumulative and progressive. If the rate of free radical production proceeds unchecked, eventually the body's natural defenses are overwhelmed. Once the containment threshold is breached, an explosive chain reaction occurs, increasing free radical concentration by up to a millionfold.

The resulting cell destruction, malignant mutation, and damage to enzymes lead to the whole spectrum of circulatory, malignant, inflammatory, and immunologic disorders that underlie the vast majority of age-related illness. The link between free radical activity and malignant change was recognized decades ago and is currently the basis for much anticancer research. The latest work of experimental pathologists suggests the major chronic diseases of Western civilization, including cancer, arthritis, senility, atherosclerosis, coronary heart disease, and related circulatory disorders, may actually be related to a form of radiation sickness, the result of continuous internal reactions, analogous to those produced by nuclear radiation.

This "China Syndrome" of human pathology can lead to a "meltdown" of cellular power plants, leading to accelerated aging and premature death. This meltdown can be largely prevented by adequate antioxidant protection from free radicals, much as the control rods in a nuclear reactor keep the rate of energy release from reaching the meltdown stage.

What has all this to do with chelation therapy? EDTA—the substance intravenously infused during chelation treatments—can reduce the production of pathological free radicals by up to a millionfold. It's not possible for free radical pathology to be catalytically accelerated by metallic ions in the presence of EDTA. Unbound metallic catalysts must be present for uncontrolled free radical proliferation to take place in living tissues. EDTA binds such ionic metal catalysts, making them chemically inert and removing them from the body. And once gone, they stay gone for quite some time.

Chelation therapy stops excessive free radical production in its tracks, halts the development of free radical disease, and allows the body to repair the damage it has already suffered. The time required for that healing process explains why full benefit from chelation therapy often takes months after a course of therapy is completed.

For the individual suffering from atherosclerosis, this means that chelation will curb the abnormal process damaging the arteries, allowing them to heal. Free radicals in blood vessels cause mutation of normal cells to atheroma cells (cells that are benignly tumorous) and promote spasm and clot (thrombosis). Free radicals block production of the hormone prostacyclin, allowing unopposed activity of another hormone, thromboxane, promoting spasm and blood clots in arteries. Red blood cells are trapped in the process. They rupture and release free iron from hemoglobin, a potent catalyst of lipid peroxidation that can increase the nearby rate of the reaction by a millionfold. This chain reaction can be interrupted by EDTA, which sweeps up and removes the unwanted metal catalysts, allowing the atheromatous plaque to heal. (More on free radicals as they pertain to atherosclerosis in chapter 17, a technical chapter explaining the scientific rationale.)

How long does chelation take?

Length of treatment varies, of course, with each patient. Chelation is normally undertaken on an outpatient basis, necessitating repeated visits to the doctor's office or clinic. Each treatment takes three hours or longer. The EDTA solution is infused as a very slow drip into a vein on the arm. Patients sit back comfortably in lounge chairs, free to chat, read, nap, or watch television.

A complete course of chelation treatment usually comprises thirty or more visits, depending upon the condition being treated. In severe cases, from fifty to one hundred treatments have been given. Frequency of treatment and dosages are tailored to individual kidney function and the patient's ability to safely eliminate EDTA in the urine. Kidney functions are closely monitored during therapy to ensure against overload. All medicines may cause harm if given in too high a dose or too rapidly. EDTA is no exception. Properly administered, the risk is very low.

"Getting chelated" is quite different from other forms of medical treatment. More about that in chapter 12, "The Chelation Experience."

5

To Be—or Not to Be— Chelated, What Every Heart Patient Should Know

Robert D., of North Carolina, has taken more than his share of lumps. As a prisoner of war in the Philippines, he survived the Bataan death march plus four years of captivity. Upon his release in 1945, Robert weighed less than one hundred pounds. What was left of him was a mess. The first thing doctors did was a gastrectomy; 80 percent of his stomach was removed.

Over the years Robert was never really healthy. Minor ailments plagued him constantly. Then he began experiencing chest pains of angina. His doctors counseled all the traditional heart ailment strategies.

"First, they advised I change my diet," Robert recalls. "Next they prescribed drugs. After that they did an arteriogram and then, finding my main arteries blocked, they said bypass surgery was the only answer."

"We'll get the very best," his family insisted.

Robert's brother arranged a surgery appointment for him at the Texas Heart Institute in Houston. His surgeon was to be the illustrious Dr. Denton Cooley.

"Wait a minute," Robert D. said. "I want another opinion."

Robert, who was associate dean at a university, is an unusually thoughtful, deliberate man with a lifelong habit of turning to books and source materials for information needed to make an informed decision.

"I didn't relish the idea of open-heart surgery," he said. "Who does? But more than that, the operation itself made little sense. What good is bypassing one small segment of an artery when the rest of the circulatory system is undoubtedly affected as well? Since plaque buildup is the end product of a biochemical process, I wondered whether there was a chemical antidote."

After further research, Robert heard about chelation therapy. It was at this point that I met him for the first time. He had decided to postpone bypass surgery over the objections of his various doctors.

"If anyone recommends chelation therapy to you," one of them had said, "don't you listen. It's dangerous—very dangerous. It can kill you."

But, as a university professor, Robert was amused by that remark. "I couldn't understand how chelation could be more dangerous than bypass surgery," he said. "Of my several friends who've had bypasses, one died on the operating table, one has required repeated surgery, and the third tells me if he had it to do all over, he wouldn't."

Within weeks after starting chelation therapy under my care, Professor Robert D. was convinced he had made the right choice.

"Being stubborn paid off," he said. "I'm breathing easier, walking straighter, feeling no chest pain, and am mentally sharper than ever."

After observing them closely for many years, I am convinced that chelation patients are not like other patients. They tend to be tough-minded individualists. Like the professor, most of them are independent and strong-willed, sufficiently self-confident to stick with their convictions despite considerable pressure to conform. (Many of them, like Robert, opt for chelation against their doctor's advice, not to mention that of well-meaning relatives and friends.)

The past twenty years have spawned both a new brand of medicine—variously called alternative, complementary, or holistic—and a new breed of patients—informed shoppers.

People are more medically sophisticated, and growing numbers now approach the medical marketplace as wary consumers. When a practitioner recommends a treatment, they want to know what they are buying.

Not nearly as reluctant as their parents to challenge doctors, many have discarded the traditional submissive patient stance in favor of dealing with their physicians on a more equal footing. They no longer see the physician as a God-like authority figure, and they insist on an active voice in their care and recovery. I, for one, welcome this new trend.

As I've previously mentioned, I'm accustomed to facing knowledgeable, well-read men and women. Many come from rural provinces but are not at all provincial. Most patients I see are well informed about their condition and the alternatives available to them. Nevertheless, I routinely block out an hour for all new patients, enough time for a constructive dialogue about their health problems and an objective discussion of their various options. And that's after my staff has documented a thorough medical history.

I assume that my readers are equally inquisitive and equally eager to be given the facts. But even men and women knowledgeable about cardiovascular disease, with better than average comprehension of the circulatory system, seldom know all that is necessary to make an informed treatment decision. Atherosclerosis is not a simple subject.

If you think you are sufficiently informed, you might save time by skipping this chapter. But before you do, take the following eight-statement test.

No fair guessing. When you've chosen an answer, you should also know why. If you get eight out of eight correct, wonderful. You can skim through the chapter. If not, it's required reading.

1. Breathing difficulties, chest pain, or leg cramps are most often associated with atherosclerosis. True or false?

2. The cause of heart attacks is always blocked circulation caused by atherosclerotic plaque. True or false?

3. "Hardening" of the arteries is a natural physiological consequence of aging. True or false?

4. Substituting margarine for butter and giving up eggs are effective ways to prevent atherosclerosis and heart attacks. True or false?

5. Low-cholesterol diets reduce the risk of atherosclerosis. True or false?

6. Vigorous sports and physical activity will protect one from developing atherosclerosis. True or false?

7. Hardening of the arteries begins with the buildup of calcium deposits to form plaque. True or false?

8. Chelation removes abnormal deposits of calcium from arterial plaque. True or false?

Ready to discover how well you did?
Every one of the above statements is false.
If you did not get every one of them right, now is the best time to learn not just how chelation works, but how your body

works, too. Do not wait until you are under the pressure of time—racked with pain or facing surgery—to find out what you need to know.

Let's go back over the test, explaining the why behind the false statements.

1. *Breathing difficulties, chest pains, or leg cramps are most often associated with atherosclerosis.*

False. You can have advanced atherosclerosis without these expected symptoms. On the other hand, you can have chest pain and not have heart disease. While exertion-related chest pain, leg cramps, or shortness of breath are the common symptoms of atherosclerosis and coronary artery disease, many other conditions can cause those same symptoms.

Gallbladder disease, hiatal hernia, and shingles are three ailments in which the pain can be remarkably similar to that of heart disease.

Cardiospasm, resulting from an irritation to the lower esophagus, is sometimes hard to distinguish from true angina.

Chest pain can stem from inflammation of rib cartilage or from arthritis in the joints between the ribs and spine or sternum. There is a condition known as nerve root syndrome (pressure or irritation of a nerve root as it exits from the spine, perhaps resulting from strain or "whiplash" injury) that can cause chest pains or anginalike pains in the arms.

Less usual, but still worth mentioning, are chest pains resulting from unsuspected broken ribs brought on by a bout of coughing. Food poisoning, or sleeping with arms or shoulders in an unnatural position, or excessive air swallowing can also cause chest pains.

Aerophagics (or "air swallowers"), much like people who hyperventilate or have overbreathing habits, are often victims of anxiety and emotional tension. Aerophagia and hyperventilation may produce symptoms similar to those of serious circulatory ailments. Depression, nervous upset, and other psychological traumas can intensify symptoms that mimic heart disease.

Then there are the many people who have experienced unsuspected "silent" heart attacks. Until a cardiogram, often taken during a routine physical exam, uncovers evidence of a past coronary event, such individuals often assume they have simply had a bad case of indigestion or an arm or neck muscle cramp. Sometimes they recall no symptoms at all.

Typical symptoms of circulatory problems that people frequently do not associate with atherosclerosis are fingers and toes that are often cold, gradual or transient memory loss, and impotence.

No one should decide on his own whether he or some member of his family has atherosclerosis or a heart problem. But a careful doctor's examination and special tests can usually determine the presence of coronary artery disease, or else the real cause of symptoms that cause concern.

So when you become nervous about your cardiac functioning, do not assume that it is due to heart disease or that it is not.

2. *The cause of heart attacks is always blocked circulation caused by atherosclerotic plaque.*

False. The degree of arterial blockage has never been completely correlated with either heart attacks or symptoms of coronary artery disease, although mechanical blockage is one important factor.

Heart attacks—called myocardial infarctions in medical jargon—may be triggered either by a mechanical blockage (an embolism, a clot, or plaque) that causes complete cutoff of the blood supply to a portion of the heart muscle, or by a coronary artery spasm, which also causes sudden blood stoppage, or by both events occurring simultaneously—with spasm superimposed on preexisting plaque. In each case, malfunction results from compromised blood flow and reduced oxygen supply. Heart muscle cells are damaged, often leading to an irreversible infarction—a scarred, nonfunctioning area of heart muscle.

Angina pains, sometimes leading to a myocardial infarction, are now considered most often to be the result of spasm superimposed on preexisting partial plaque blockage. Heart attacks, however, have been well documented in the complete absence of plaque, caused entirely by spasm of the muscular arterial wall in a perfectly normal artery.

Conclusive proof that coronary spasm without plaque can lead to coronary thrombosis and myocardial infarction has been described in detail. In the July 28, 1983, issue of the *New England Journal of Medicine,* a report was published of a twenty-nine-year-old woman who had a history of anginal-type chest pain. Coronary artery ateriograms were performed, and the first injection of dye showed no plaque and a completely normal left coronary artery, without any evidence of atherosclerosis or other mechanical obstruction. During a second injection of dye, a few minutes later, a diffuse spasm was seen in the left anterior descending artery. Under direct x-ray visualization, a clot was observed to form in the area of spasm and the patient rapidly developed a full-blown myocardial infarction with an area of muscle death. (The doctors kept her heart beating and the patient lived.) This report documents the first time when the entire process from spasm through thrombosis and subsequent heart muscle death took place with physicians watching. And it occurred in a normal coronary artery with no plaque whatsoever.

It now appears that in a surprisingly significant number of cases, coronary artery spasm or primary metabolic failure of the myocardium (heart muscle) triggers a heart attack in the relative absence of atherosclerosis. Numerous earlier reports of normal arteriograms in patients with myocardial infarctions support the conclusion that spasm without atherosclerosis is not that uncommon.

Conversely, there have been numerous published reports of individuals with complete blockage of all three main coronary arteries who had no abnormal heart symptoms at all (an American astronaut, for one example), no angina, and excellent tolerance to strenuous physical exertion. Such cases are found at autopsy following accidental death or death from causes other than cardiovascular disease. When three astronauts perished in the tragic Project Apollo launch-pad fire on January 27, 1967, postmortem examinations revealed that all three of these men, supposedly in superb health, showed signs of atherosclerotic disease—one with so advanced a case he would have been a bypass surgery candidate had he ever experienced symptoms. Although he was running more than ten miles daily and passing semi-annual astronaut physical examinations, all three of his coronary arteries were severely blocked.

The free radical shutdown of prostacyclin production can cause unopposed spasm of blood vessels and an increase in platelet accumulation, leading to clotting. This explains why angina or myocardial infarction frequently occurs after a fatty meal replete with oxidized fatty acids that lead to an explosion of free radical activity. Free radical pathology now appears to be an important factor.

While most people are familiar with dangers to the heart posed by plaque accumulations inside the arterial walls, all too few are aware of the equal or greater danger from arterial spasm. More to the point, there is a general lack of public awareness as to what causes arterial spasm and the preventive measures that might ward it off.

Improper calcium/magnesium ratios within the arterial muscle cells appear to be one important contributing factor leading to arterial spasm. The heart muscle is also unable to pump rhythmically and adequately when intracellular magnesium is low and intracellular calcium is high.

A proper ratio of calcium outside cells to magnesium inside cells must exist in order for proper contraction and relaxation of muscle cells to occur. This is true for both the heart pumping muscles and the muscle fibers that encircle the arteries and control blood flow.

Calcium (as well as sodium) is in higher concentration outside muscle cells, and magnesium (as well as potassium) is more concentrated within cells, more so when the muscle is at rest. For contraction of muscle fibers to occur, there must be a partial reversal of the calcium and magnesium ratios with calcium entering the cells and magnesium leaving the cells. Sodium tends to parallel the movement of calcium, and potassium moves in concert with magnesium. To complete the contraction-relaxation cycle, the prior concentration gradients must be restored. This requires a pumping action energized by oxygen. If circulation is compromised, and oxygen supply is reduced, calcium and sodium pumping is slowed, resulting in impaired relaxation and arterial spasm.

The vital question then is: How does the calcium/magnesium ratio get out of whack? Best evidence suggests that it occurs partially as a result of generalized nutritional imbalances and inadequate magnesium in the typical Western diet. (More about this important topic later.) Ongoing free radical damage also causes calcium to leak through cell membranes.

3. "Hardening" of the arteries is a natural physiological consequence of aging.

False. Many people have the mistaken idea that as one gets older, one's arteries "harden" as though after a limited number of years, the artery's inherent destiny is to become rigid, blocked, and inflexible.

Not so. Hardening of the arteries, while most prevalent in the elderly, is the result of metabolic changes that may actually start early in life. Early stages of the process leading to formation of atheromas (benign arterial tumors), with hardening of the arteries caused by cross-linkages and calcium deposition may begin at a young age. To some extent, this can be attributed to a diet of highly refined foods (white flour, white sugar, white rice) commonly consumed in industrialized nations. That type of diet is deficient in the many vitamins, minerals, and antioxidants necessary to protect against oxygen radical damage.

4. Substituting margarine for butter and giving up eggs are effective ways to prevent atherosclerosis and prevent heart attacks.

False. Margarine and egg substitute manufacturers have successfully popularized a massive public misconception: that a low-cholesterol diet lowers blood cholesterol because of a decreased consumption of cholesterol, when it's more likely the reduction in harmful fats that turns the trick. One study even showed an increase in heart disease among people who used margarine on a regular basis. They were half again as likely to suffer a heart attack, compared to people who did not consume margarine and other hydrogenated fats.

Cholesterol is necessary for many vital functions such as the emulsification, digestion, and absorption of fats, the synthesis of sex hormones, the production of vitamin D via an interaction with sunlight on the skin, and antioxidant protection of the lipid membrane surrounding every cell. Cholesterol not only enters the body as food, but is predominantly manufactured internally by the liver and other organs.

Two-thirds of all the body's cholesterol is produced by the body's own cells and does not come from dietary cholesterol intake. Much of the cholesterol deposited in atherosclerotic plaque is produced within the body and is not derived from food cholesterol directly. Some is produced within the plaque itself. What is not widely known is that cholesterol is harmless and even beneficial if it is not oxidized. It's essential for life and health in its natural form. Only when oxidized does cholesterol cause harm. Antioxidant supplementation is key to protect cholesterol from this damage. And EDTA chelation therapy removes undesirable metals that greatly speed cholesterol oxidation.

5. Low-cholesterol diets reduce the risk of atherosclerosis.

False. This statement is true only if you reduce the total amount of fats consumed—not just the cholesterol content. As stated before, most cholesterol in the blood is produced by the body and does not come from dietary sources. As a matter of fact, cutting out cholesterol can be of no benefit, since the less cholesterol is consumed, the more is produced internally by the liver.

The now-popular dietary recommendation to consume less cholesterol may be somewhat beneficial but for another reason. The main benefit comes from the fact that a low-cholesterol diet will usually result in less total fat consumption. Fats are easily oxidized and, in excess, can overwhelm the body's antioxidant defenses. Average Americans eat 45 percent of their calories as fat. Reduction to approximately 30 percent of caloric intake as fat, combined with antioxidant supplementation, will greatly reduce that risk factor. Many fats consumed in food are already oxidized in the cooking process and thus speed release of oxygen radicals in the body.

It is not cholesterol, per se, that is the culprit. Most people suffering from high serum cholesterol, and who are at risk of atherosclerosis, are endangering their health by consuming too much fat—not from eating too many high-cholesterol foods. In my opinion, laboratory norms are set far too low. Blood cholesterol levels up to 280 or even 300 can be normal if the HDL (good kind) cholesterol is normal.

As a matter of fact, the normal population can add up to three eggs a day to their diet with no increase whatsoever in blood cholesterol, provided those eggs are soft boiled, hard boiled, poached, or steamed—not fried or scrambled in fat.

In controlled studies conducted at Rockefeller University, two-thirds of subjects eating three eggs daily had only a slight increase in blood cholesterol. Their bodies responded to the dietary challenge with an automatic cholesterol regulation process, which tends to keep blood levels constant.

It seems clear from experimental evidence that diets that contain less fat do reduce the risk of atherosclerosis.

Less than 1 percent of the population has an inherited tendency to dangerously high blood cholesterol (400 mg/dL or higher), and these remarks do not apply to them. Victims of that genetic trait should follow a very strict diet and must often take prescription drugs under medical supervision to lower blood cholesterol.

A recently published review of the world's medical literature showed no correlation between blood cholesterol levels and extent of atherosclerosis observed at autopsy or during surgery. The only patients with a significant correlation were the rare few with an inherited genetic trait for extremely high cholesterol levels—above 400 mg/dL.

6. *Vigorous sports and physical activity will protect one from developing atherosclerosis.*

False. Autopsies performed on highly active people, including marathon runners, have disclosed far advanced atherosclerosis with extensive arterial plaque buildup, although these people rarely exhibited symptoms of disease. Similar investigations of highly mobile populations, such as the nomadic members of the Masai tribes in Africa who routinely walk between twenty and thirty miles a day, have also uncovered a surprisingly high incidence of atherosclerosis in a people long thought to be free of this disease. The Masai diet is high in peroxidized fat and oxidized cholesterol.

That is not to suggest that regular exercise is without benefit, or that devotion to an exercise program is a waste of time. The physically active lifestyle of the Masai is a good example of the protective effect of exercise. Like other superactive people, the Masai (while not immune to plaque buildup) rarely suffer symptoms of arterial blockage from this condition. It seems that physical exercise has a protective effect, preventing arterial occlusion by promoting collateral circulation around the blocked arteries, and by causing compensatory enlargement of the plaque-filled arteries. While exercising has not been proven to stop plaque buildup, it does improve the ratio of HDL (high-density lipoprotein) cholesterol to total cholesterol. HDL cholesterol is the unoxidized portion and has a retarding effect on disease development. HDL cholesterol has potent antioxidant traits itself.

Individuals who maintain high levels of physical activity have many fewer symptoms of atherosclerosis, and fewer atherosclerotic-related deaths, even when they in fact have extensive arterial plaque. And they feel better, with more energy, increased mental alertness, and less depression.

Indeed, other published research also suggests that vigorous exercise reduces the risk of sudden death from coronary heart disease. When a research team at the University of Washington in Seattle and the University of North Carolina in Chapel Hill evaluated activity levels among 1,250 persons who died suddenly and unexpectedly of heart disease, they found sudden-death victims had participated less in high-intensity leisure time exercise—such as jogging, chopping wood, swimming in a pool, tennis, or squash—than a matched group, who were more active and had a lower sudden-death incidence.

In another study, it was found that bus drivers who sat all day had a much higher heart attack rate than did bus conductors who were on their feet all day, running up and down steps on double-decker buses. Furthermore, when Dr. Ruth K. Peters of the University of California studied 2,779 Los Angeles policemen and firemen, she found the incidence of heart attack can be dramatically lowered with twenty or thirty minutes of vigorous exercise three or four times a week. It seems that exercise does help to protect one from the harmful effects of atherosclerosis; but it does not prevent the disease from developing and laying the groundwork for the harm it will eventually cause—perhaps, in the case of the physically active person, in old age rather than middle age.

7. Hardening of the arteries begins with buildup of calcium deposits to form plaque.

False. The initial event in arterial disease is damage to the arterial lining, resulting from blood flow stress, routine wear and tear, or free radical damage. To understand the basic concepts underlying the development of atherosclerosis, visualize these cells within the arterial walls abnormally multiplying until they form a benign tumor or scar, a growth akin to the cellular proliferation seen in cancer. These arterial wall cells mutate in response to free radical damage to the genes contained in the nucleus, identical to the way in which atomic radiation causes mutations. An occasional cell loses its ability to control cell division and multiplication, resulting in uncontrolled, tumorlike growth. Growth-promoting factors that speed plaque enlargement are also released.

The tumor (atheroma) thus formed is nonmalignant and will not metastasize or spread to other parts of the body. But it is nonetheless an unwelcome, space-occupying mass on the inside of the artery, which accumulates collagen, elastin, and other connective tissue constituents, eventually growing to block the flow of blood.

When this growth exceeds its blood supply of oxygen and

nutrients, it begins to break down in the center, becoming decayed or necrotic, and gradually gathers deposits of cholesterol and calcium. As it grows into what we call plaque, it becomes progressively firmer and more rigid. Calcification is actually a late occurrence in plaque formation—not the initial event.

8. *Chelation removes abnormal deposits of calcium from arterial plaque.*

False. Before the free radical concept of degenerative disease was known, it was tempting to hypothesize that EDTA chelation had its major effect on calcium metabolism. We now know that calcium is just another link in the chain of cause and effect.

EDTA can influence calcium in many ways, but the calcium-chelation connection has been blown way out of proportion and has been a major weak link in past explanations of how and why EDTA is a beneficial treatment.

I believe that one of the primary reasons why chelation has been accepted so reluctantly by the medical profession at large is past reliance of chelation advocates—lacking any other justification for observed improvement in patients—on unsubstantiated "roto-rooter" types of explanations, which have destroyed credibility with knowledgeable scientists.

Newly emerging scientific discoveries now enable us to demythicize the calcium-chelation connection, so we shall devote the next chapter to more thorough investigation of this entire issue.

6

THE CALCIUM-CHELATION MISCONCEPTION

Let us clear up a major—and damaging—misconception about chelation therapy. It does not strip calcium out of plaque or calcified "hardened" arteries.

In the early days, chelation therapy pioneers, eager to explain the treatment in terms patients could grasp, described EDTA's action as "pulling rivets out of a bridge" or "de-cementing the lining of blood vessels." These were crude but vivid ways of describing what they thought was happening—that calcium was being dislodged and drained from "hardened" and plaque-clogged arteries.

Ever since that time, chelation advocates have often stressed the removal of calcium as the underlying reason for chelation "payoffs." This is both scientifically inaccurate and, to a large degree, responsible for delaying universal acceptance of chelation therapy.

Why then have doctors stuck with the calcium-chelation connection when it is highly hypothetical, nebulous, and vulnerable to the justifiable criticism of knowledgeable biochemists and cardiovascular physiologists?

The use of chelation therapy with EDTA for atherosclerosis was not discovered or developed in a research lab, a teaching hospital, or a university medical school. It evolved in doctors' offices—a suspect setting in the view of the majority of physicians who have far more respect for new modalities that emerge from academia.

Some physicians in private clinical practice were trained many years ago, prior to the introduction of intensive biochemistry in medical school. They are far more interested in helping patients regain health than in delving into the complex biochemical pathways responsible for therapeutic effects. The notion that calcium removal shrinks plaque was a convenient explanation that fit best with what was known at the time.

Until recently, most published scientific references to EDTA, relating to either industrial or medical applications, concerned its effect on calcium. EDTA's ability to remove calcium has freed consumers from "ring around the bathtub." Internal medicine specialists value it in that regard, knowing it to be useful for lowering calcium in patients with life-threatening high blood calcium. The FDA-approved package insert specifies EDTA's use as a calcium chelator for hypercalcemia, caused by advanced metastasized bone-dissolving cancers, and digitalis toxicity (potentiated by calcium).

Degenerative diseases have long been linked to calcium-related excesses. It's well established that as people age, they accumulate calcium deposits in unwanted places, especially in arteries and arterial plaque. EDTA's proven calcium-lowering ability, coupled with demonstrable improvement in the condition of patients so treated, led to the seemingly reasonable conclusion that "zapping" calcium out of the hardened areas was the way EDTA worked.

With 20/20 hindsight, it's easy to ridicule so simplistic a notion. Although EDTA is certainly a calcium chelator, its affinity for calcium is far less than for many other metals, such as iron, copper, lead, mercury, cadmium, and aluminum.

EDTA will quickly release calcium to bind with ions of one of those other metals and will transport them out of the body through the kidneys. Serious investigators, noting that the amount of calcium removed by one infusion of EDTA is less than half a gram—not much more than would normally be excreted by the body in one day—remained rightly skeptical.

The explanation satisfied the uninitiated, but made chelation appear flaky to the sophisticated, and discredited not only the proponents of the calcium-chelation connection concept, but the therapy as well. Thus the baby was thrown out with the bath water. Not believing the tale of how it worked, many well-trained professionals chose to dismiss tales of how well it worked as equally fanciful.

After more than twenty years, the most expedient route to bringing chelation into the mainstream of medicine might be to bury it along with its tortuous history and begin afresh, to "rediscover" and rename the entire process in light of current new knowledge. I have no doubt, were this to happen, that chelation therapy would, under its new name, be hailed overnight as the medical breakthrough the world has been waiting for.

Since that does not seem feasible, our best alternative is to replace misconceptions with accurate, scientifically supportable, updated theories of what EDTA does and does not do. There probably is a calcium-chelation connection, but it's secondary to EDTA's impact on free radicals and trace elements. I now believe that the free radical-chelation connection is very important to explain improvement in individuals with atherosclerosis, disordered calcium metabolism, and degenerative diseases in general.

If EDTA does not simply "pull out calcium rivets" or "decalcify" plaque, how then does it reverse symptoms of atherosclerotic disease?

To summarize what is currently known about the cause and effect relationships in atherosclerosis: The initiating event in hardening of the arteries is a localized injury to the lining of a blood vessel wall. This superficial loss of cells occurs repeatedly during normal daily activities from the following four causes:

1. Ongoing "injuries" of a low magnitude are a normal result of the stress of blood flow and routine wear and tear. Such localized injuries are more frequent and severe in the presence of high blood pressure. Healthy defenses rapidly heal these small defects.

2. Free radicals cause damage to blood vessel walls and LDL cholesterol. Once oxidized by free radicals, LDL cholesterol becomes toxic to blood vessel walls. Thus, atherosclerosis is speeded by the presence of free radicals, which have the ability to proliferate, producing a cascade of other free radicals—a case of a poison producing more poison.

3. Ongoing blood vessel injury may be immunologic. That is, the body's immune system, which is essential for proper resistance to disease-causing organisms and environmental contaminants, may inappropriately attack healthy cells.

4. Recent evidence indicates that chlamydia, a common bacterial infection, may also contribute to the initial injury.

For a variety of reasons, related to excessive free radicals and trace element imbalances, a blood vessel injury can result in a tumorlike atheroma, rather than normal healing. The plaque, which has no blood vessels within, continues to grow until it becomes so large that adequate oxygen and nutrients cannot diffuse into its core from the surrounding blood. The center of the plaque then degenerates, eventually picking up deposits of calcium and cholesterol. Calcification is a relatively late event in the development of arteriosclerosis—hardening of the arteries and hardening of the plaque with deposited calcium.

While there are quite a number of nutritional and lifestyle factors that contribute to the formation of atherosclerotic plaque and its calcification, we shall deal here specifically with the form of disordered calcium metabolism most directly corrected by EDTA.

Long before abnormal calcium deposits become solidified and visible on x-rays or to the naked eye at autopsy, there is a progressive fine mist of calcium building within cells and tissues. The absolute amount of calcium within the body is not the most important factor. Abnormalities first occur with calcium distribution in various places—such as bone versus soft tissue and in the ratio of calcium to magnesium.

An optimal calcium-to-magnesium ratio results in a much higher concentration of calcium outside the cells than within. Healthy cells, with active energetic metabolism, are able to maintain an efficient pumping system within their lipid membrane walls that keeps calcium out and magnesium in.

Calcium outside of cells is normally ten thousand times more concentrated than within the cells. It therefore takes a lot of energy and an efficient pumping mechanism to keep the calcium outside of cell walls, against such a high gradient. This is analogous to the dikes of Holland, a country below sea level, which must both hold the water back and constantly pump it out to prevent flooding. If calcium is allowed to increase within cells, it causes damage and cell death.

Magnesium is just the opposite. It is kept mainly within the cells.

With age, those concentration gradients deteriorate and slowly reverse. Calcium levels increase within the cells as magnesium levels decrease, partly because of gradual breakdown of the cell's pumping mechanism, which can be damaged by many things, including free radicals. Also, free radical damage to the cell wall creates "leaks" that allow calcium to seep in and magnesium to seep out, in other words, similar to holes in the dike that leak faster than the pumps can compensate for. A healthy cell wall, by contrast, is quite impermeable to these metal ions.

The more calcium leaks into a cell, the more poisoned its metabolism becomes, and eventually concretelike deposits form, causing it to die. The more cells that die, the fewer are left to keep the basic life processes going.

Excess intracellular calcium and an abnormally high calcium-magnesium ratio speed cell death, especially when blood flow and oxygen delivery become compromised by atherosclerosis. Arterial spasm becomes much more intense and calcium leakage into cells proceeds much more rapidly once the pump is weakened by a disrupted cellular calcium-magnesium ratio.

A recent study shows that other essential trace elements—including zinc, cobalt, chromium, and iron—also leak into cells when blood flow is compromised.

A vicious cycle is set in motion as diminished blood flow and reduced oxygen speed leakage into cells. Intracellular calcium causes muscle cells surrounding arteries to go into more intense spasm, acting like a tourniquet, causing further reduction in blood flow and oxygenation, speeding calcium influx, and on and on. Other trace elements that leak through can also poison cells.

This self-perpetuating process underlies the success of recently introduced calcium "blockers," such as nifedipine and verapamil, which slow the entry of calcium into muscle cells. The calcium "blockers" work by an entirely different mechanism than EDTA, and they have no lasting effect on the underlying disease process.

Impairment of the calcium-magnesium pump also allows more calcium in the ionized state to enter the cell (as distinct from protein-bound calcium), activating enzymatic production of prostaglandin-related leukotrienes. That chemical reaction releases more free radicals.

Leukotrienes are potent inflammatory substances that attract white blood cells to the area. White blood cells produce free radicals as "bullets" to attack foreign invaders. When excessively stimulated by leukotrienes, white blood cells run amok and initiate excess free radical production, leading to increasing inflammatory damage to healthy tissues. Small blood vessels dilate, causing swelling, edema, and leakage of red blood cells and platelets through blood vessel walls, which result in micro-thrombi (microscopic clots). Some red blood cells then hemolyze, releasing free copper and iron, which in turn catalyze an explosive increase of free radical destruction to lipid membranes in the vicinity by up to a millionfold, triggering a further vicious cycle.

The accumulation of intracellular and connective tissue calcium is further speeded by abnormal vitamin D activity of certain cholesterol oxidation products created nearby by free radical reactions. This vitamin D-like activity can produce a form of localized vitamin D toxicity in cells and tissues, the effect of which is to further speed calcium deposits.

Once free radical production exceeds the body's threshold to protect itself, metabolic breakdown, cellular damage, and tissue calcification all take place more rapidly, even exponentially. It now appears that calcium deposition is just another link in the chain of cause and effect created by free radical damage. EDTA may influence and speed removal of calcium deposits in many ways as part of the healing process, but not as directly as formerly hypothesized.

EDTA does briefly reduce blood calcium levels during infusion and hastens excretion, but if you look at the one million grams of total calcium in the body, the one-third gram excreted is extremely small—far less than dietary calcium intake on an average day and barely more than normal daily urinary excretion. That cannot be the main benefit of the therapy.

During the short time that EDTA is circulating in the body (it has a half-life of less than one hour), it temporarily lowers blood calcium. The resulting drop in serum calcium provides a stimulus to the parathyroid gland to step up production of parathormone. This hormone, in turn, signals osteoblasts in bone to increase their production of normal bone calcification, drawing on other calcium sources in the body, some, presumably, from pathological deposits. This pulsed, intermittent parathormone stimulation, produced by each EDTA, is known to cause a lasting effect on these osteoblasts, of approximately three months' duration. This is a proven effect of EDTA. It is theoretically possible that this parathormone effect contributes to benefits following chelation therapy. That is one theory.

It has been widely reported over the years that the greatest results following EDTA chelation therapy occur approximately ninety days or more after the completion of a course of treatment. This corresponds to the duration of increased bone uptake of calcium by osteoblasts, but also may also reflect a more important EDTA effect-reduction of free radical damage allowing gradual healing, which also takes months to become evident.

If, as the newest research indicates, the free radical theory of degenerative disease is correct, then preventing and reversing free radical pathology would be very important in the treatment and prevention of such major age-related ailments as atherosclerosis. By removing metallic catalysts from the body, EDTA chelation is unique in its ability to control free radical reactions in a lasting way and also to improve metabolic health by removing other toxic metals, some of which are not related to free radicals.

Specifically, EDTA reduces the rate of localized pathological free radical chemical reactions by up to a millionfold, below the level at which the body's defenses can once again take over, and thus provides time for damage to be repaired by natural healing.

Let's review the physiologic effects of EDTA in relation to free radical pathology:

• EDTA removes these excess metal ions, which accumulate with age, and permits normal healing of peroxidative damage after stopping further damage.

• EDTA controls free radical damage. Free radicals cause cell destruction by initiating chain reactions of lipid peroxidation, which disrupt fatty membranes within and surrounding cells. Lipid peroxidation is greatly speeded by the presence of abnormally located metal ions such as excess iron.

• EDTA removes lead, aluminum, and other poisonous metals, restoring normal enzyme functioning. Excessive concentrations of heavy metals poison cells and cause enzymatic dysfunctions independent of their action as free radical catalysts, with a concomitant loss of cellular homeostasis and failure of vital cell functioning.

• Abnormally high concentrations of essential nutritional trace elements in blood-deprived tissues can also act as metabolic poisons. EDTA also acts to remove and redistribute those elements.

• EDTA enhances the integrity of cell membranes by controlling oxidative damage. The ability of cells to produce and store energy depends on healthy membranes surrounding each cellular compartment, including each tiny, interior energy factory, called a mitochondrion. EDTA is known to stabilize mitochondrial membranes and to enhance the efficiency of energy metabolism, independent of any effect on blood flow or oxygen.

• EDTA helps reestablish prostaglandin hormone balance, which in turn reduces arterial spasm, blood clots, plaque formation, and arthritis. Prostaglandins are extremely potent hormones, with a half-life measured in seconds, and must be constantly synthesized. The two most important prostaglandins, in relation to blood vessels and atherosclerosis, are prostacyclin and thromboxane. The first reduces the tendency of platelets to stick and cause blood clots and reverses blood vessel spasms. The second does just the opposite. It causes intense spasm and stimulates platelets to become sticky, converting blood vessel walls to "fly paper." In fact, a proper balance between the two must be maintained to protect against injury and hemorrhage on the one hand and to maintain normal circulation on the other. Prostaglandins are produced from fatty acids, and production imbalanced by fatty acid peroxidation. EDTA inhibits spread of such lipid peroxidation by chelating out the catalyzing metallic ions.

• EDTA protects the integrity of blood platelets. Platelets are tiny, cell-like corpuscles in the blood stream that quickly adhere to areas of injury and release substances that diminish blood loss by forming clots and promoting healing. Those same growth factors speed plaque growth. In the process of clotting, platelets change shape, tend to become very "sticky," and attach themselves to the walls of diseased coronary, cerebral, or other arteries. Platelets also release spasm-producing thromboxane. After treatment with EDTA, platelets lose their tendency toward overcoagulation and spasm is reduced.

• EDTA sweeps up minute molecules of polyvalent metals such as abnormally situated iron and other metals, which accumulate abnormally through the body as it ages. Concentrations of metallic ions that have the ability to speed free radical chain reactions are so tiny that even the minuscule residue in distilled water can initiate such reactions. The addition of EDTA in the laboratory completely blocks the free radical chain reactions that would otherwise occur, by binding metallic ionic catalysts, making them chemically inert.

• EDTA normalizes calcium metabolism by reactivation of enzymes poisoned by lead and other toxic heavy metals. Most important, EDTA blocks the free radical mediated conversion of cholesterol into substances with excessive, localized vitamin D activity—activity that causes plaque to accumulate calcium more rapidly.

• EDTA intermittently lowers serum calcium, allowing gradual reversal of abnormal calcification, by stimulating uptake of calcium in bones. EDTA has been proven not to cause osteoporosis, and the evidence shows that EDTA may even help to reverse osteoporosis.

• EDTA increases tissue flexibility by uncoupling age-related cross-linkages. Cross-connections between large protein and other connective tissue molecules cause increasing rigidity and loss of flexibility with aging. Cross-linkages between enzyme proteins can block normal biochemical activity. The metals removed by EDTA can link between molecules and cause progressive loss of elasticity, flexibility, and metabolic function.

• EDTA encourages overall metabolic efficiency due to the whole range of factors given above.

The development of atherosclerosis, or any other degenerative disease, takes decades and is almost always the result of not one, but a combination of health-destroying factors. In like manner, reversing the process cannot rely on just one "magic bullet." Impaired health can only be restored by therapies that counteract a spectrum of disease-causing factors.

Underlying every known risk factor contributing to age-related degenerative diseases—from atherosclerosis to cancer—is excess free radical activity with resulting damage.

Improper diet causes free radical damage; so do smoking, abuse of alcohol and drugs, insufficient exercise, excessive stress, and lack of adequate antioxidants contained in vitamins, minerals, and trace elements. The key to restoring health is to limit excessive free radical production by adopting health-promoting lifestyle strategies, daily use of nutritional supplements, and, if need be, more directly and quickly by a course of chelation treatments. Many of my patients take chelation therapy as a preventive measure, long before symptoms of ill health occur.

With our new understanding of free radical chemical reactions, there is now a sound basis for universal acceptance of EDTA chelation therapy.

7

FIRST, THE GOOD NEWS: OTHER CHELATION PAYOFFS

Is it really true that chelation improves memory, reduces insulin requirements in diabetes, restores sight, increases sexual potency, lessens the aches and pains of rheumatoid arthritis, smoothes away wrinkles, reverses and slows senility associated with Alzheimer's disease, and ups longevity?

For many years, it's been downright embarrassing to me, a Harvard Medical School graduate and past president of my county medical society, to be in the position of endorsing a treatment that sometimes makes me sound like an itinerant medicine man drumming up sales for a suspect new cure-all.

While I have never represented chelation to be a treatment for any conditions other than those stemming directly from atherosclerosis, it is impossible to ignore the other benefits experienced by chelation patients. Of course, placebo effect plays a role here, just as in all other types of therapy. But benefits following chelation are lasting, while placebo effects tend to fade rapidly.

One of the great drawbacks to gaining acceptance for chelation has been its literature, history, and lore, crammed as they are, with reports that chelation is successful not only in reversing the symptoms of atherosclerosis and circulatory blockage, but in improving patients' health status in many other, less predictable, ways.

Case in point: Bessie B.

When seventy-year old Bessie first came to the clinic, she presented a laundry list of health complaints typical of people her age who suffer from atherosclerotic cardiovascular disease.

"There's so much wrong with me, I don't know where to start," she said.

Bessie's daughter, Flora, who had all but carried her feeble mother from the car to the office, filled in the medical details. Bessie's past clinical history included two heart attacks, chronic pulmonary disease in the form of bronchial asthma and emphysema, severe osteoarthritis, heart failure, chest and leg pains, bone softening (osteoporosis), cataracts in both eyes, insomnia, and lately, mental confusion characterized by increasing forgetfulness.

"Sometimes, I don't even remember Flora's last name," Bessie cried.

"After Dad died," her daughter related, "Mom was okay for awhile, but then she got to the point where she couldn't do for herself. She couldn't remember to take her medicine as directed. Sometimes she'd skip it; other times she'd double-up and overdose. She had difficulty expressing herself. She was terribly depressed, and she talked about dizzy spells. If she walked upstairs she'd get terrible chest and leg pains. Some days she forgot to eat, and we were always afraid she'd walk off and forget how to get home.

"We've had Mom everywhere, but it seems that all the medications and drugs have just added to her problems. Each time they prescribe something that helps one condition, it brings on another.

"Every doctor I've seen has told me Mom will never be able to look after herself or live alone anymore. The last one said it would all be downhill and that eventually she'd require nursing home care. Do you want to know what he told me? He said, 'Don't look so glum. By that time, she won't know the difference.'"

Many doctors do not like to treat old people. It's frustrating because you can't "cure" old age. But Flora wasn't ready to give up on her mother. When her neighbor, a nurse, told her about chelation, she hoped for a reprieve.

Fifteen weeks and thirty treatments after Bessie started chelation therapy, she resumed housekeeping, able to care for herself once more. Better still, Bessie celebrated Thanksgiving by cooking dinner—turkey and all the trimmings—for a houseful of relatives. There was a family reunion that Bessie's kinfolk, twenty in number, are still talking about.

"Some of my folks hadn't seen me in a while, and they couldn't imagine me this much improved," Bessie chuckled. "I just never get tired of hearing them marvel at how well I look."

A case of macular degeneration: Mrs. Ormal D.

Ormal was eighty-two years of age. She was gradually losing her sight but was otherwise remarkably healthy for a woman her age. She was bitter at having to curtail her activities and depressed at the prospect of becoming totally blind.

"I love to read, do needlework, paint, watch TV, and visit people. I like to walk, garden, and keep busy," she told me.

"Now, the doctors have told me I have to learn to live like a blind lady. I'll try anything—diet, vitamins, chelation, you name it—that will help me retain some degree of sight."

Ormal's medical records confirmed severe macular degeneration, involving problems with central vision in both eyes caused by degenerative changes in retinal blood vessels on the back surface of the eye. By the time we began chelation treatments, almost all vision was gone from her right eye. Her ophthalmologist had known of other patients suffering with similar macular degeneration who had been dramatically helped by chelation therapy, and so he referred Ormal to me.

"I just held my breath the first few treatments," said Ormal. "I couldn't see a thing out of my right eye, and the other was getting worse. Then, after my fifth treatment, I was at home lying on the sofa. For some reason I raised myself up, closed my left eye, and looked out toward the front door and caught sight of the sky and a neighbor's terrier crossing the lawn.

"I couldn't believe it! I was seeing and with my bad eye. I jumped up and ran outdoors, holding my breath for fear it wasn't true. But when I looked all around, and saw the trees and everything okay. I started yelling 'I can see! I can see!' Let me tell you, I stirred up the neighborhood that day."

It wasn't long before Ormal was sewing a bit and reading again. She can even find names in the telephone directory now. The last time she visited me for a follow-up chelation she had just returned from a trip to India and showed off pictures of herself riding on an elephant. It's unusual for chelation to work that quickly and dramatically, but in my experience, there's no treatment that works as well for macular degeneration.

Without exception, chelation physicians have noted that patients enjoy multiple benefits. Most experience symptom relief not only from the one specific ailment for which treatment was undertaken, but also from other, secondary complaints—lessening of aches and pains, joint stiffness, dizziness, ringing in the ears, mental sluggishness, and general fatigue.

Occasionally, there are rapid, dramatic results, such as a recovery from acute paralysis, reversal of blindness, or a restoration of hearing. Rex S. could not use the telephone without an amplifier prior to chelation; now he can. Bill D., who received thirty-five chelations following a severe heart attack, was astonished to find his gray hair, especially on his chest, gradually turning dark once again. Bill's greatly impaired cardiac ejection fraction, measured by noninvasive isotope gated blood flow testing, more than doubled following chelation; it increased from 20 percent to 40 percent.

Another sixty-seven-year-old male told me he was able to resume normal sexual function after many years of difficulty. He had been a victim of Peyronie's disease, which is characterized by the formation of hardened scar tissue on one side of the penis causing an erection to be bent at an angle, making intercourse painful or impossible. After thirty-five chelations, the scar tissue dissolved completely, and a normal erection was again possible.

At first exposure, many of those recoveries seem unbelievable, but to chelation specialists, who have witnessed scores of "can you top this?" case histories, they have become routine. The published literature also contains many case histories documenting that chelation frequently pays off in unexpected ways. I'm afraid that if I related all the stories my patients tell me about improvements following chelation, I'd lose my credibility.

As early as 1963, a report in the *Medical World News* stated that patients with diabetes had reduced insulin requirements after chelation, and periodic infusions of EDTA helped keep complications of their disease under control. According to this medical periodical, heart specialists Dr. Lawrence E. Meltzer and Dr. J. Frederick Kitchell were "excited to discover" that chelation with EDTA offered very positive benefits in diabetes, and that some patients were able to discontinue insulin injections. (It has been my experience that, after chelation, insulin requirements are usually reduced by approximately half.)

In addition, the doctors reported, some patients with diabetes suffering with severe peripheral vascular problems, who had lower extremity gangrene, were saved from undergoing leg amputations. One such patient, Dr. Kitchell recalled, "ended up walking the entire seven miles of the Atlantic City boardwalk, and couldn't walk a block before we started therapy." (For the record, for unknown reasons and despite very favorable data, Kitchell and Meltzer later swerved away from chelation. See chapter 10 for a detailed discussion.)

While physicians were uncertain as to the specific ways in which chelation improved the course of diabetes, they speculated it had something to do with EDTA's ability to alter enzyme activity that controls blood sugar, perhaps by segregating and removing toxic metals and by restoring a balance to the essential minerals and trace elements. Iron, zinc, copper, cobalt, and manganese were five metals named as possibly contributing to the pathology of enzyme systems in diabetes. (We now know that iron, copper, and manganese in excess can act as potent free radical catalysts.)

When their studies showed that patients with diabetes excreted abnormal patterns of metallic elements in their urine during chelation, this finding added credibility to their theory. Zinc is involved in the proper functioning of many separate enzyme systems, and it also functions as part of the storage mechanism of insulin.

Abnormal metal excretion patterns, by the way, have been noted not only in diabetes, but also in rheumatoid arthritis patients and in individuals with cancer. This supports recent conclusions that trace metal imbalances that can interfere with normal enzyme functioning may in turn lead to increased free radical activity and speed the development of many serious degenerative ailments. Metals are involved in virtually every metabolic pathway, and there are other theoretical explanations for benefit, in addition to the free radical theory.

A research study published in 1999 by Dr. Frustaci and colleagues in Italy showed that when the heart muscle was starved for circulation in coronary artery disease, intracellular levels of many metals increased in heart muscle cells. Of the essential nutritional metals, zinc increased by 300 percent, cobalt increased 600 percent, chromium increased 800 percent, and iron increased 500 percent. Such increases are potentially toxic, and chelation with EDTA could correct them.

Even the essential, nutritional metallic trace elements are toxic in excess. Such excesses of intracellular concentrations poison enzymes and increase pathological free radical reactions. Of the solely toxic metals, even at very low levels, antimony increased 600 percent, and mercury increased 400 percent. EDTA has little effect on mercury, but that toxin can easily and inexpensively be removed from the body with DMSA (dimercaptosuccinic acid, taken by mouth—available generically from compounding pharmacists and from any pharmacy as succimer or Chemet).

EDTA acts not just to remove metals from the body, but also to restore a balance, rearranging essential elements from locations of excess to other places in the body, reinstating a more natural distribution. Perhaps that is how chelation causes its major benefit. There are many theories and we are still not sure which is most important.

According to John Olwin, emeritus clinical professor of surgery at Rush Medical College at the University of Illinois, "It seems reasonable to assume that chelation with EDTA, by removing some of the more than 50 contaminating trace metals that have been found to accumulate within the human body, may revitalize enzyme systems damaged by their presence."

Dr. Olwin, whose speculation preceded many years of current knowledge of the way in which trace metals catalyze free radical activity, was impressed by clinical results with chelation, although he admitted he was unable to fully explain how specific improvements came about. Nevertheless, he could see no reason why the therapy's incomprehensibility should stand in the way of patients enjoying benefits similar to those he had observed. He began using chelation many years ago, first on patients suffering from obliterative atherosclerosis in their legs. Most of them were candidates for summary amputation.

"We saved many of those limbs," Dr. Olwin has noted. "The necrotic areas were limited; they healed; the limbs warmed; the nails and hair began to grow."

Olwin extended treatment to patients suffering from a variety of circulatory diseases, and, along with relief from pain, he reported the healing of ischemic ulcers, a decrease in recurrent thrombophlebitis, and improvement in mental processes.

"In the early stages," reported Dr. Olwin, "we noted that people with gangrenous limbs and gangrenous toes, in some instances were not as mentally sharp as they had been. One executive who had lost his job because of this returned to work after he had been treated with EDTA.

"Some patients note an increased nail and hair growth, and say they've been getting haircuts more often than they did before.

"There's been an increase in libido and potency reported by some patients. One man, a once brilliant, 80-year-old lawyer, had had chronic brain syndrome (senility) for many years and then was chelated. About a year later, his wife called one day to ask, 'Is this supposed to increase sexual desire?' When she was told it's been reported to have that effect, she replied, 'Well, James hasn't been interested in sex for 15 years, and last night he tried to make love to me.'"

Erectile failure—the most persistent and troublesome male sexual complaint—is often related to atherosclerosis, which may severely restrict the blood supply to the lower limbs and pelvic area. Danish psychologist Dr. Gorm Wagner, a specialist in human sexuality research, has linked both impotency and frigidity to constricted blood flow, basing his conclusions on studies that used a penis-linked camera to photograph changes in the sex organs during simulated intercourse.

If there is a circulatory component to some forms of Parkinson's syndrome, as many specialists believe, that would help explain the reported instances in which sufferers of this ailment have improved after chelation therapy. Although no clinical studies have been done to date, oxygen starvation affects the same parts of the brain as the encephalitis virus infection that in later years can lead to Parkinson's disease.

Another possible mechanism by which chelation might reduce symptoms could be the removal of aluminum from body tissues. Autopsies of Parkinson's syndrome patients have revealed higher than normal concentrations of aluminum in the affected parts of their brains. Aluminum causes cross-linkages in free radical damaged tissues. Dr. Harman believes that free radical damage-induced oxidation of dopamine receptors is the true cause of one form of Parkinson's. Improvements in symptoms of Parkinson's disease following chelation are not as consistent as seen in patients with proven circulatory disorders, but they occur often enough to warrant further research.

What about the aches and pains of rheumatoid arthritis? Why do patients so often report less discomfort after chelation?

Rheumatoid arthritis is actually a collagen (connective tis-sue) disease, and chelation is known to restore collagen to a healthier state by reducing cross-linkages, not just those caused by metallic ions, such as calcium and aluminum, but also those associated with the increased linking of sulfur atoms on adjacent coils of springlike connective-tissue molecules.

Incidentally, the FDA has approved an oral chelating agent called d-penicillamine for use in cases of rheumatoid arthritis. This drug chelates copper, lead, and other trace elements and, although similar to EDTA in some of its actions, is far more toxic. And d-penicillamine does not reverse atherosclerosis.

There are high hopes that oral chelates may be the wave of the future, but thus far there is no evidence of benefit in age-related and atherosclerotic problems. Unfortunately, EDTA is so poorly absorbed by mouth that it remains in the digestive tract and blocks absorption of vital nutrients. EDTA must be given intravenously for benefit.

But do we have an explanation for Bessie's recovery from "incurable" senility?

Once a relatively rare problem, "senility" has become epidemic. It is estimated that there are currently more than two million senile persons in the United States. Some leading psychiatrists are predicting that more than five million Americans will be similarly afflicted by the year 2020. By age eighty-five, approximately half of all Americans suffer some form of Alzheimer's dementia.

Not all cases of "senility" are related to vascular disease. True dementia—medically defined as the deterioration of or the loss of intellectual faculties, reasoning power, and memory, and usually characterized by confusion, disorientation, apathy, and some degree of stupor—can also be the result of a virus or prion infection, as in the rare (and always fatal) disorder called Creutzfeldt-Jakob disease, or of a genetic disorder such as Huntington's disease. Dementia can also occur as a result of Parkinson's disease, multiple sclerosis, and stroke.

There is also so-called multi-infarct dementia, caused by multiple little strokes caused by ischemia (blockage of blood supply) to numerous small areas of the brain. These can also be caused by "showers" of debris released from ulcerated plaques in the larger arteries of the neck. Loss of brain function following stroke or injury has recently been reported to improve with hyperbaric oxygen therapy.

More than half of all so-called dementias, however, are caused by Alzheimer's disease, which results in a progressive deterioration of brain function. Alzheimer's disease, which tends to strike its victims in middle age, results in a relentless form of "brain rot" generally considered to be irreversible.

No less an authority than Dr. Neal R. Cutler, former chief of the section on brain aging and dementia of the Laboratory of Neurosciences at the National Institute of Aging, has been quoted as considering Alzheimer's disease "incurable."

Alzheimer's disease extracts a tremendous price. Caring for an Alzheimer patient is a nightmare. He or she is typically paranoid, suspicious, delusional, hostile, irritable, and irrational, insisting on performing tasks such as tending financial matters, or driving the family car long after becoming unable to do so.

Eventually, overburdened families admit they are unable to cope. More than half of all admissions to nursing homes are Alzheimer victims, and their care costs an average of thirty thousand dollars a year or more.

But is Alzheimer's as hopeless as most scientists claim? Is senility an inevitable consequence of our living longer?

Chelation therapists have valid reasons for a more optimistic outlook. Most now believe Alzheimer's is treatable and, in some cases, at least partially reversible. Not only is there clinical evidence to support that view, but also scientific support in the new findings linking trace metal accumulations to Alzheimer-type mental deterioration.

One neurological phenomenon of Alzheimer's disease is the finding of tangled fibers disrupting brain cells on microscopic examination of victims' brains. Dr. Daniel P. Perl of the University of Vermont has found abnormal accumulations of aluminum within the affected brain cells of those neurofibrillary tangles.

Going one step further, a team of researchers from the National Institute of Neurological and Communicative Disorders and Stroke (NINCDS), headed by Dr. D. Carleton Gajdusek, found high accumulations of aluminum in the brains of Chamorro natives of Guam who had died of either amyotrophic lateral sclerosis (ALS) or Parkinsonism dementia.

This population has been closely watched because it has been adversely affected by a high incidence of those two chronic disorders, both of which were previously suspected to be transmitted by a slow-acting virus. For several years, Dr. Gajdusek and his colleagues have been studying the implications of the high levels of aluminum found on the island of Guam.

Because of the similarities between the Parkinsonism dementia syndrome and the presenile form of Alzheimer's dementia, other researchers have kept a watchful eye on these studies. In both disorders, there is an excessive accumulation of aluminum-containing neurofibrillary tangles in the brains of victims. The newest findings confirm Dr. Perl's earliest work correlating high concentrations of aluminum in the brain with the development of these tangles.

Furthermore, NINCDS scientists are exploring the possibility that mild and subclinical hyperparathyroidism caused by dietary imbalances of calcium, magnesium, and phosphorous could contribute to accumulations of toxic concentrations of aluminum, especially in specific areas of the brain.

These are important discoveries because they confirm what chelation physicians have long been claiming: chronic "subclinical" environmental metal toxicity is one of our important health problems. Toxic metals, such as lead, can be shown to depress our immune systems as well as our brain functions. This breakdown of immunity is related to the onset of cancer and also causes allergies.

Dr. Carl Pfeiffer, former director of the Brain Bio Center in Princeton, New Jersey, has also researched blood aluminum levels and has found a very high correlation between loss of memory and elevations of blood aluminum.

If aluminum is a contributing factor in Alzheimer's, little wonder this disease has grown to epidemic proportions, affecting more and more people at younger and younger ages. Aluminum, one of the most common elements in the earth's crust, has no known function in our body, but we are absorbing more and more of it through our digestive tracts.

While relatively large amounts of aluminum have always been present in food, there is reason to believe we now get a heavier dose than our bodies can comfortably handle.

Our early ancestors ingested a significant amount of aluminum from the rocks and stones that they used to mill grain into flour. But we have added aluminum to our diets in great quantities. It is not only in all our junk foods and our drugs, but it is also added to our water supply by treatment plants. It gets into our foods via the use of aluminum pots and pans, the wrapping of acid foods such as tomatoes in aluminum foil, and the storing of foods in aluminum cans.

Aluminum is a common component of many of today's foods. They include processed cheeses (aluminum is added as an emulsifying agent), pickles (as a firming agent), baking soda, and cake mixes (as a leavening agent). It is used in the manufacture of antacids, antiperspirants, buffered aspirin, and vaginal douches. It is added as a drying agent to table salt so that it will "pour when it rains."

Subclinical aluminum poisoning appears to be a serious potential, and generally unsuspected, problem. Antacid tablets, for instance, commonly contain 200 mg or more of elemental aluminum, with a recommended dosage of up to twenty-four tablets daily, only ten of which would put the user more than one hundred times over the presumably acceptable

20 mg a day mean. Similarly, buffered aspirin contains 10-52 mg of aluminum per tablet; a fourteen-tablet-per-day regimen would also result in high aluminum intake. More and more aluminum is infiltrating our environment, some from cooking utensils and some in the form of acid rain, which dissolves aluminum from the soil.

The body has historically had good defenses against aluminum intoxication by preventing absorption from the digestive tract. There is now evidence that common dietary imbalances of calcium, magnesium, and phosphorous, with a synthetic form of vitamin D added to so-called fortified foods, have resulted in an increase of aluminum absorption from the intestines. Acid rain has resulted in a greater uptake of aluminum by food plants.

When aluminum gets into the bloodstream, it may adversely affect the parathyroid gland, causing a further imbalance of calcium-regulating hormones, which in turn further increases the absorption of ingested aluminum. This self-perpetuating cycle results in higher levels of tissue aluminum than have occurred in past generations. Aluminum can cause cross-linkages between collagen and elastic tissue molecules, increasing rigidity or sclerosis of soft tissues and speeding that aspect of aging.

Is it purely coincidental that we are witnessing a devastating increase in the incidence of all chronic degenerative diseases at the very same time that exposure to potentially toxic metals takes place at higher levels than ever before?

Quite unlikely, especially since chelation—known to remove unwanted metallic elements from the body—has proven remarkably effective for the very ailments it is now suspected are caused, at least in part, by metal poisoning.

When Richard Casdorph, M.D., Ph.D., assistant clinical professor of medicine at the University of California Medical School in Irvine, an internist and cardiovascular specialist, published a research study of the efficacy of EDTA chelation therapy in brain disorders, he documented a measurable increase in brain blood flow in all but one of fifteen patients. The statistical probability that this could have been due to random chance or a placebo effect is less than one in ten thousand.

Using radioisotopes to measure before and after chelation cerebral blood flow, Dr. Casdorph was able to show for the first time that chelation is unquestionably useful in the treatment of senility where impaired cerebral blood flow is a factor.

One of Dr. Casdorph's subjects was a seventy-six-year-old white female, clinically diagnosed with Alzheimer's disease, with CT scan evidence of cerebral atrophy (brain shrinkage). At the onset of therapy, she was confused. After twenty infusions of EDTA, there was "marked improvement in the brain blood flow study, as well as significant improvement in her mental functioning," Dr. Casdorph reported.

Another patient, a seventy-two-year-old lady with a long-standing history of documented cerebral atrophy, was having delusions and hallucinations and at times did not recognize her husband of fifty years. On more than one occasion, after walking out to the sidewalk in front of her home, she had not been able to find her way back into the house. Before Dr. Casdorph included her in his study, her husband was considering institutionalizing her. "After the first six infusions (of EDTA chelation therapy), all of the above mentioned symptoms cleared, the patient was completely oriented and rational," Dr. Casdorph writes. "It was no longer necessary to consider institutionalization. There also occurred some improvement in her vision."

In Dr. Casdorph's view, when toxic metals have accumulated over the years, they impair enzyme reactions and block metabolic pathways, and thus may accelerate the development of age-related degenerative diseases. Chelation, which acts only on a spectrum of metals and removes the toxins, lead, cadmium, and aluminum, from the body, may improve cell function.

Canadian investigators have also now found aluminum at above normal levels in the brains of some Alzheimer's victims, and they speculate that senile dementia might more properly be called aluminum dementia. Clinching the metal poisoning tie to Alzheimer's is evidence from the University of Toronto that chelating metals from the body either halts or reverses mental deterioration in Alzheimer's syndrome.

When Dr. Donald McLaughlan, professor of physiology and medicine of the University of Toronto, attempted a small clinical trial with an iron chelating agent (deferoxamine), six treated Alzheimer's patients improved, while eleven untreated controls did not, as judged by the Wechsler Intelligence and Memory Scale, signal detection tasks, and an EEG (electroencephalograph). Deferoxamine is most potent as an iron chelator (iron is a free radical catalyst known to accumulate with age) and binds less tightly to aluminum. If Dr. McLaughlan had used EDTA, I suspect his results would have been even better since brain cells are highly sensitive to free radical damage. EDTA acts in a similar way to deferoxamine, but has a much broader effect.

At Ohio State University, investigators have taken a different approach. They have embarked on a five-year study of Alzheimer's patients to determine the long-term effects of low-aluminum diets in conjunction with agents that limit aluminum absorption.

Incidentally, Dr. Johan Bjorksten, a prominent gerontological researcher, pointed out that aluminum is building up not just in our brains, but in all our tissues, particularly in our blood vessels, such as the aorta. According to Dr. Bjorksten, aluminum potentiates the formation of cross-linkages, causing blood vessels to become stiff and "hard," like old, dried-out garden hoses. Indeed, he has discovered a relationship between aluminum accumulation in tissues and death rates of the population at large. Aluminum levels are significantly lower in tissues of individuals who are active and aware in their eighties and nineties as compared with matched populations exhibiting dementia and severe ravages of aging.

As previously noted, the original use of EDTA in the United States was for the treatment of lead toxicity. It was applied to battery factory workers suffering from lead poisoning, a rare occurrence more than thirty years ago, but not so unusual today.

Lead poisoning has become a major health threat. It has been established that we have five hundred to seven hundred times more lead in our bones than our ancestors did. Those lead deposits, it appears, are released from bone into our bloodstream under stress, following a severe trauma, or during a high fever from an infection, thus poisoning us most when our resistance is lowest.

No one is immune. The affluent person as well as the ghetto resident is at risk, as Dr. Vernon Houk, director of environmental health services at the Center for Disease Control in Atlanta pointed out. He said, "Lead toxicity is not, as once thought, confined to lower-income urban areas."

Lead is absorbed by eating, breathing, or touching contaminated surfaces. In previous years, concern primarily centered on lead ingestion via paint or industrial exposure. Today we have identified dozens of hazardous sources—atmospheric lead from smelters, coal, and leaded gasoline from motor vehicle exhausts. Leaded gasoline accounted for half or more of the lead in our bodies a decade ago, although it is now less of a problem since the advent of unleaded gasoline.

If you eat food grown near the roadside, drink milk or eat meat from animals grazing on our lead-contaminated pastures, or cook in improperly glazed pottery, you increase your exposure to lead. Moreover, there is absorbable lead in a bewildering variety of seemingly innocuous products: mascara, tobacco, curtain weights, newsprint, toothpaste, hair dyes, canned fruit and fruit juices, wine (with leaded caps), electronic circuit boards, and some pesticides. In recent years there has been a trend to use less lead, but lead pollution persists in the environment.

Subclinical exposure can result in subtle, yet significant, adverse health effects. In children, it manifests itself in behavioral problems, particularly hyperactivity and learning disabilities. In adults, relatively low levels may trigger headaches, digestive disorders, and general irritability.

Lead and aluminum, unlike the nutritional elements, have no "normal" site to bind them. They are always toxic.

If chelation did nothing more than eliminate lead from the body—and EDTA is in fact the recognized treatment of choice for this very problem—it would deserve the equivalent of a medical "Oscar."

But EDTA does much more than just get the lead out. It may be the answer to another of our most pressing health problems: reducing the incidence of cancer.

One piece of significant research, which has thus far been ignored, was published in the *Journal of Advancement in Medicine* in 1989 by a Swiss physician, Dr. W. Blumer, and myself.

Mortality from cancer decreased by 90 percent during an eighteen-year follow-up of fifty-nine patients treated with EDTA chelation therapy. Only one of the fifty-nine treated subjects (1.7 percent) died of cancer whereas 17.6 percent of nontreated, matched control subjects died of cancer. Statistical analysis was highly significant—the probability that this was pure coincidence is less than two in ten thousand. That study was reviewed by the medical faculty at the University of Zurich. They could not find any flaws in these findings. The conclusion? Chelation was solely responsible for a 90 percent decrease in the death rate from cancer during the eighteen-year follow-up. As for deaths from all other causes—cardiovascular disease included—overall mortality rates were significantly lower in the chelated group.

If this benefit is confirmed by subsequent research, it could be one of the most important advances in our fight against cancer. Thus far, however, these findings have been totally ignored by cancer researchers. EDTA chelation therapy remains "not politically correct" in academic research circles, regardless of published benefits.

I must stress here that the reported benefit against cancer was preventive, and that patients in this study all received EDTA chelation before any cancer had developed. This study shows that chelation might reduce cancer by up to 90 percent if received prior to the onset of malignancy. There is yet no evidence that chelation is of benefit once cancer has been diagnosed.

Dr. Harry B. Demopoulos, associate professor of pathology at New York University Medical Center and an internationally known cancer researcher, has also correlated heavy metal accumulation, free radical activity, and lipid peroxidation with the initiation and promotion of cancer.

Research from Japan introduces an entirely new anti-cancer aspect of EDTA. It was found that EDTA injected intravenously into mice increased the blood concentration of interferon four- to twelvefold. Interferon is produced by the immune system. This might explain why some patients feel so much better immediately upon being chelated. They could be suffering impaired immunity that is relieved by EDTA and increased interferon production.

Chelating physicians cautiously avoid claims that chelation reverses the aging process, both because it is too difficult, complex, and expensive to prove and because it might serve to further alienate our nonchelating colleagues.

One graphic example, however, of the relationship between uncontrolled free radical activity and accelerated aging is the genetic disease known as progeria, caused by the hereditary absence of free radical scavenging enzymes. TV viewers may remember having seen two victims of this heart-wrenching ailment some time ago when the cameras trailed them on a visit to Disneyland. Though neither had yet celebrated a tenth birthday, they resembled little old men. Typical progeria sufferers are wrinkled, have dried and sagging skin, are bald, bent, and have frail bodies crippled by arthritis and advanced cardiovascular disease before puberty. They usually die of "old age" while still in their teens. One form of progeria has been successfully treated by administering the free radical scavenging enzyme peroxidase.

Our biological time clock, which begins ticking at birth, seems to run at a speed determined by our body's efficiency in taking command of free radical chemical reactions. Reduced defenses against destructive free radical activity accelerate the aging process. Further evidence comes from the animal kingdom. Studies show that those mammals with the longest life spans have the highest relative levels of SOD (superoxide dismutase), an anti-free radical enzyme produced within the body.

As noted before, EDTA not only restricts free radical proliferation, but also allows the body's natural antioxidant defenses to regain control. Both are good reasons to accept EDTA chelation as a credible approach to healthful life extension.

No discussion of possible benefits of chelation therapy can be complete without mention of its potential for life extension. Research going on for more than three decades has continually demonstrated that the life spans of some very primitive organisms are dramatically extended when chelating agents are added to their culture medium. In some tests, increases of up to 50 percent the normal life span were recorded.

When distinguished gerontological researcher Dr. Johan Bjorksten reviewed and analyzed all the pertinent research and interpreted the data as it might apply to humans, he estimated that the immediate widespread use of chelation therapy would result in average life span increases of from nine to eighteen years for males and from seven to sixteen years for females.

To sum up the good news: We can expect chelation therapy to result in dramatic health improvements that go far beyond those linked directly to improved circulation. Bringing free radical reactions under control buys time—time for ailing cells to rebuild, time for rejuvenated cells to repair damaged tissue, time for diseased organs to heal.

Dr. Bjorksten considered it reasonable to hope for an ultimate 15 percent increase in life span when men and women are routinely chelated preventively early in life. Whether you are presently fifteen—or far older—being chelated offers you renewed hope for a substantially expanded high-quality life span.

8

NOW, THE BAD NEWS: YOU'LL HAVE TO FOOT THE BILL

As of this writing, being chelated is much like having extraordinary beauty or great wealth—a rare and distinctive privilege to be enjoyed only by a relatively small, elite portion of the population.

If you are not related to a physician (many doctors are "closet chelators" who routinely chelate themselves and their loved ones but do not offer this treatment to patients), therapy is available only to those who can afford to pay for it—three thousand to four thousand dollars or more—out of their own pockets.

Even if you have the most comprehensive medical insurance coverage and thought yourself protected from every possible major or minor health problem, it is no go. Neither government nor private health insurance carriers will pay for the medical expenses of chelation therapy, except in rare instances.

When I say rare instances, I mean exactly that. Only a very few times in my many years of experience with thousands of chelation patients have health insurers paid a claim. I remember them well, for in almost every instance, payment was some kind of fluke.

For example, take the case of Mr. M., a patient who owned a business and paid large monthly medical insurance premiums for his more than three hundred employees. When he developed coronary heart disease and opted for chelation therapy, his company's insurance carrier turned down his request for payment.

He didn't argue.

"No problem," he smiled. "I'll just cancel our company policy and take the company business elsewhere."

His check arrived in the mail that week.

Then there was Mr. D., a man who held a senior position with a large, well-known health insurance firm, a company which by the way has routinely turned down chelation patients' claims. When this gentleman required chelating, having decided it was his best chance for recovery from a heart attack, he had no difficulty getting his claims paid. Easy for him. He authorized the signing of his own check!

And then there was the case of Mrs. K., a lovely lady, gentle and soft-spoken, but without such well-placed connections. Not knowing any better, she mailed her bills for chelation treatment off to Medicare and promptly received a check for $1,600 by return mail. When she told me, I was dumbfounded.

"Oh my," she enthused at her next visit, quite pleased. "You see, doctor, they do pay after all."

Two months later, she sent off a packet of additional bills, for her continuing series of chelation treatments including all the proper substantiating documents, and back came the more usual Medicare response: "We don't pay for that."

"But, you paid last time," she complained.

Obviously, some red-faced bureaucrat had goofed—and all she has received since have been demands to give the $1,600 back.

On a few occasions, when their labor union leader applied leverage to the insurance carrier and was willing to go to bat for them, union members have been reimbursed. And claims are sometimes paid by smaller, more obscure local-type insurance companies. It's my impression they are so far out of the mainstream, they don't know what chelation is.

That just about covers the exceptional circumstances under which you might have your chelation bills paid. You must be someone, know someone, or be gosh darn lucky. So be warned, the bad news about chelation is it is only going to hurt you in one place—the pocketbook.

You do have one more option. You can sue! And you'll probably win.

Litigious patients have on almost every occasion won judgments against insurance companies when they have been denied payment on frivolous grounds.

Courts have typically taken a dim view of attempts by insurance companies to deny a policyholder's payment for whatever medical treatment he prefers, as witness the judicial opinion written by Judge Francis N. Pecora of the City Court of New York City in the case of Brigitta Bruell against Associated Hospital Service of New York and United Medical Service, Inc.

Ruling against the health insurance companies that had refused payment to Mrs. Bruell because the treatment she selected was not FDA approved, the judge wrote: "Nowhere in the contract is the word 'necessary' treatment defined as a treatment which must be FDA approved. Moreover, implicit in contracts of this nature is the notion that a patient has a right to rely on his physician's decision as to what types of treatment are necessary for use against the patient's ill conditions.

"If the Defendant's [the insurance company's] position is upheld, which permits them to make value judgments as to correctness of judgment of duly licensed physicians, then the rights of the patients to such contracts will become highly subjective and will be determined at the whim and caprice of Defendants' medical advisory committee, whoever they may be.

"If medical coverage were to be denied each time an insurance company could provide a licensed physician to testify that, in his expert opinion, a treatment given by another licensed physician was not medically necessary or efficient, then no medical treatment could survive a denial of coverage."

In a similar case, Judge Paul Jones of Industrial Claims in Brevard County, Florida, ruled that the bills of Robert J. Rogers, M.D., be paid by the carrier, Continental National American Insurance, on behalf of the employee, Mr. Gerald Tillman.

Mr. Tillman, prior to consulting Dr. Rogers, had received extensive medical treatment, including several surgical procedures for reduction of back and leg pains, but continued to suffer. Only after receiving chelation therapy did he experience pain relief. Continental, however, denied payment. Judge Jones ruled for Mr. Tillman, stating Dr. Rogers's treatment of the patient with chelation therapy was "both reasonable and necessary and the employer/carrier should be responsible for the charges for treatment of the claimant in accordance with the standard medical fee schedule."

When the state of Florida, in a separate legal action, then attempted to restrict Dr. Rogers's chelation practice, the District Court of Appeals ruled that the State Board of Medical Examiners was "without authority to deprive physicians' patients of their voluntary election to receive chelation therapy in treatment of arteriosclerosis simply because that mode of treatment had not received endorsement of the majority of the medical profession."

After lengthy hearings, Chief Judge Boyer admonished those who had attempted to squash Dr. Rogers's practice of chelation, saying for the record: "History teaches us that virtually all progress in science and medicine has been accomplished as a result of the courageous efforts of those members of the profession willing to pursue their theories in the face of tremendous odds, despite the criticism of fellow practitioners.

"Copernicus was thought to be a heretic when he theorized that the earth was not the center of the universe. Banishment and prison was the reward for discovering that the world was round. Pasteur was ridiculed for his theory that unseen organisms caused infection. Freud met only resistance and derision in pioneering the field of psychiatry.

"We can only wonder what would have been the condition of the world today and the field of medicine in particular had those in the midstream of their profession been permitted to prohibit continued treatment and thereby impede progress in those and other fields of science and the healing arts."

The courts once again spoke out in defense of chelation when Medicare refused to pay claims for *any* type of treatment for *any* patient who consulted Dr. Leo J. Bolles (a state of Washington medical doctor) for *any* reason, on the grounds that he offered services "not consistent with good medical practice." Dr. Bolles included chelation therapy in his practice.

After two years of hearings on EDTA chelation therapy, at which I and many other experts testified, it was Judge Gordon McLean Callow's opinion that "The record is replete with examples of dramatic improvement in some patients as a result of this unique strategy."

Commenting on the negative testimony about chelation that had been presented by the antichelation physicians, the judge noted a "decided reluctance on the part of organized or established medicine to undertake further research on chelation therapy. In this writer's opinion, this reluctance borders, based on the evidence in this record, upon medical negligence."

Judge Callow classified the long and intensively fought court battle as "another skirmish in the intense war between the medical 'establishment' and the medical 'mavericks,' for want of more polite descriptions." And expressing admiration for the "maverick" physicians who had testified in favor of chelation therapy, he offered this opinion:

"While I do not believe the undercurrent allegation that the medical establishment is against chelation therapy because of its possible impact on the current $20,000 bill attached to each heart bypass surgery, the sometimes overtly arrogant approach taken by certain of the government's witnesses in this case toward any possibility that the 'chelationists' might be on to something is only otherwise explicable on the supposition that the medical 'establishment,' including the pharmaceutical houses, are unwilling to experiment with EDTA unless huge monetary rewards await."

The judge concluded, "If these men are right in their obviously honest convictions (and they may be), the medical profession owes at least a portion of its enormous research resources to a fair and objective trial of chelation therapy in the treatment of vascular disease, obviously a major physical problem in American society."

Hopeful as these victories are for those patients who turn to the courts, there is a "Catch 22" aspect to seeking judicial relief via a breach of contract suit against your health insurer. Even if you win, you lose, inasmuch as you will most likely incur $15,000 or more in court and legal costs to secure a $3,000 to $4,000 reimbursement.

In one spectacular example, a wealthy patient, able to pursue the issue on principle, fought in the courts for almost five years, at an out-of-pocket cost of more than $15,000, before he "won" a $2,750 judgment.

The way things now stand, I advise chelation patients that they can come out ahead by not submitting their bills for chelation treatments to their insurance carriers. Once insurers get wind that bills being submitted are in any way associated with chelation, they may refuse to pay for other treatments received by that same patient elsewhere.

To make things worse, many health insurance companies have recently rewritten their contracts in a way that legally denies payment for non-FDA approved therapies and places the decision-making process about what is necessary with insurance company employees.

Hard to believe? Perhaps. But true all the same. When people learn that their health insurance will not cover chelation treatments, they are initially stunned. Then come the questions.

"Is chelation legal?"

"Why won't they pay?"

"What's going on?"

First things first. Chelation therapy is completely legal. Doctors who provide this treatment are not operating on the fringe. Nor are they violating any state, federal, or local statutes or acting contrary to the highest medical ethics. State and federal courts have repeatedly reaffirmed a licensed physician's freedom to use any procedure or drug which, in his considered judgment, is to the best interest of his patient.

The American Medical Association's position on this issue is equally clear. In response to a query on the question of the legality of chelation therapy, Dr. Asher J. Finkel (Scientific Activities Division, AMA, Chicago) wrote, "We (the AMA) have always maintained that any licensed physician is free to use any drug approved by the FDA for marketing, in any way the physician deems appropriate in his best clinical judgment. If your physician believes in the usefulness of chelation therapy, for whatever purpose in his clinical judgment, he is perfectly free to use it."

If legality is not at issue, what is?

At a time when more than 70 percent of medical expenses in this country are paid by third-party insurers, how are Medicare, Blue Cross, and other insurance companies able to refuse payment for chelation?

As in many other situations where ordinary citizens are forced into "go-fight-city-hall" wrangles with huge, impersonal organizations, when you trace the problem to the source, you usually find bungled legislation or bureaucratic distortion of legislated guidelines. So it is here.

To understand what has been happening, a brief explanation of Medicare is necessary. The federal medical insurance program for people sixty-five and older was passed into law in 1965 under the Social Security Act. It consists of two parts:

Part A: Hospital insurance covers all eligible persons and is funded by compulsory contributions out of workers' wages and payments by their employers.

Part B: Medical insurance, supplementary and optional, is available for persons sixty-five and older who pay a modest monthly premium for approximately 80 percent reimbursement based on "allowable charges" for doctor, laboratory, and surgical fees, and other medical services not included under Part A. Part B Medicare reimbursement is administered by private health insurance companies that are paid with taxpayer dollars to scrutinize payment requests and weed out the fraudulent or excessive.

When Medicare contractors refuse payment for chelation, they are following official guidelines from Washington that state that "EDTA chelation therapy is a covered service only if administered for: emergency treatment of hypercalcemia, control of ventricular arrhythmias and heart block secondary to digitalis toxicity, or (with another form of EDTA) for lead poisoning."

The guidelines were prepared with the help of government scientists and physicians who have little or no experience with chelation, or have come to an adverse judgment without a careful study of the actual hard scientific data, or are representatives of a competing specialty such as cardiovascular surgery and balloon-stent cardiology.

According to Medicare protocol, chelation therapy is not "medically indicated and necessary" for the treatment of atherosclerosis and therefore does not meet "accepted standards of medical practice." They have labeled chelation "experimental" because "there is a lack of well-designed, controlled clinical trials of its effectiveness."

Do these instructions to Medicare contractors to deny coverage for chelation therapy for atherosclerosis represent a sincere and honest attempt to protect the gullible from the danger of an unproven treatment? You be the judge.

While we shall cover the issue of EDTA's safety more thoroughly in another chapter, keep in mind that multiple studies, as well as almost five decades of clinical experience with a million or more patients, have demonstrated the treatment to be safe when properly administered.

Experimental? With more than one million people having taken more than twenty million treatments to date—almost as many as have undergone bypass surgery?

Let's consider Medicare's interpretations of "medically indicated and necessary." Imagine the response to that one from men and women, now recovered, who successfully turned to chelation as a last resort.

It is all too clear that the difficulties lie not with chelation, but with the way the Medicare statutes were written and are being interpreted. There can be no doubt that congressional intent, under the general provisions of the Medicare statute, was for the treating physician, together with his or her patient, to be the primary determinant of what is "medically indicated and necessary" treatment and care.

Congress also included in the Medicare statute a provision guaranteeing patients free choice among health service providers, indicating their steadfast resolve not to interfere with the normal physician/patient relationship, nor to allow a Medicare carrier to override the professional medical judgment made by a competent physician in good faith.

It is when you examine the "proven effectiveness" challenge that it becomes obvious that patients' health insurance rights are being violated, for Medicare and its contract administrators are demanding a higher order of proof of both safety and efficacy for chelation therapy than for many other accepted-for-payment treatments, such as coronary bypass, angioplasty, balloon angioplasty, stents, and vascular surgery.

The United States Congress has established the Office of Technological Assessment (OTA) to help the government study and assess emerging technologies. OTA was asked by Congress to evaluate the usual and customary medical practices in this country. OTA subsequently submitted a report entitled "Efficacy and Safety of Medical Technologies," which concluded with this startling statement: "Only ten to twenty percent of all procedures currently used in medical practice have been shown to be efficacious by controlled trial." Although that report is now more than a decade old, little has changed since.

Close scrutiny of the 133-page document reveals a decided antichelation bias involved in the way payment guidelines are applied. Health insurance carriers have routinely, without question, shelled out billions for treatments never subjected to controlled clinical trials and never proven beneficial.

The OTA was highly critical of (among other procedures) coronary bypass surgery, pointing out that this procedure had become the primary approach to treatment of coronary artery disease, with as many as 550,000 such surgeries now performed each year, at an annual costs of more than $25 billion, despite there having never been clearly demonstrated benefits compared to nonsurgical treatments. The report singled out coronary bypass surgery as an example of widely used technologies that "have been diffused rapidly before careful evaluation," adding, "Claims that the operation prevents death remain largely unproven."

As for procedures that had been tested by controlled trial, it was the conclusion of the OTA study that even those formal studies which had been completed and published are of questionable value, inasmuch as careful review of research reports in leading publications revealed that more than 75 percent had invalid or insupportable conclusions as a result of statistical problems alone. The conclusion was that few published clinical trials are well enough designed to yield valuable results.

"Personal experience," the congressional report suggests, "is perhaps the oldest and most common informal method of judging the efficacy and safety of a medical technology."

Precisely. Witness the personal experiences of more than one thousand chelating physicians who each and every day see the best evidence of all that they are using an effective treatment—recovered patients.

Following the government's lead, private health insurance companies have taken the stance, "What Medicare won't pay for, we won't either." Blue Cross/Blue Shield, as a matter of fact, has a published "hit list" of nonreimbursable procedures. (Hyperbaric oxygen for stroke and chelation therapy for atherosclerosis are among them.) And those decisions are heartily endorsed by mainstream professional associations, which are commonly biased against competing procedures.

"Surely," patients say, "things will eventually change. The government (and the insurance industry) can't hold out forever."

I wouldn't bet on it because it's a matter of bucks.

If the truth be known, the government can't afford the payments it's already committed to, much less take on new obligations. The looming Medicare crisis has Congress hard-pressed to find ways to cut services, not add new ones.

Let's assume for a moment that the government has a change of heart. Medicare and the health insurance industry agree to pick up the tab for chelation therapy.

The first year or so the savings would be substantial. If only one-fourth of the 550,000 persons annually undergoing vascular surgery (a $30,000+ procedure) chose chelation instead (at a cost of roughly $4,000), health insurers would save a minimum of $3 billion.

Why then so much resistance? I don't know, but perhaps they have looked further down the line. Perhaps they are not as oblivious to chelation therapy's benefits and potential as they pretend to be. As chelation experts are all too well aware, the therapy is needed by far more people than those already in so sorry a condition that they are candidates for surgery. How many more? That is the trillion-dollar question.

No one knows for sure how many potential chelation patients are out there, but we do have figures upon which we can make an educated estimate.

What percentage of our entire population has serious atherosclerosis? Over 50 percent of the deaths in this country are caused by circulatory ailments related to arterial disease. If paid-for chelation were available to just those already exhibiting symptoms (this usually happens five to ten years prior to death), we are talking about chelating about thirty million people for a total cost of $120 billion. A mind-boggling figure.

But what about all those men and women without symptoms? Autopsy studies conducted on American soldiers killed in action during the Korean and Vietnam wars revealed arterial plaque formation in the majority of these seemingly healthy young men, most in their early twenties. Improved diagnostic techniques, such as nuclear magnetic resonance, total body CAT scans, ultrafast CAT scans of the heart, digital subtraction angiography, and enhanced ultrasound imaging, have the potential to detect early atherosclerotic disease in more than half of all adult men and women (and many boys and girls over the age of puberty). Now we are talking about 100 million chelation candidates, and the cost could be equivalent to a third of the federal budget.

Is Medicare (and the health insurance industry) determined to save you or themselves? The health insurance industry is in trouble. Many are in business to make money, to pay dividends to stockholders. That is tough to do at a time when health care costs are skyrocketing faster than higher premiums can be charged. And if payments skyrocket for chelation, premiums must be increased proportionately, and they are already too expensive to be affordable for many people. As U.S. Senator Edward Kennedy once said, "To the [health] insurance companies, people are the enemy, because every private claim is a threat to corporate profit." Even so-called nonprofit medical insurers reward their top management with exorbitant salaries and fringe benefits paid for with money not spent on health care.

It is no secret that the government is already struggling with the budget, that Social Security is in deep trouble (mainly because more people are living longer than originally anticipated), and that Medicare threatens to bankrupt the system.

Government economists estimate the Medicare fund could be more than $1 trillion in the red by the year 2005. Medicare payments are rising faster than revenues in the system's Hospital Insurance Trust Fund, not only because of rising health care costs and the proliferation of medical services, but because there are growing numbers of elderly citizens enjoying longer life spans.

Unpleasant as it is to think of our government as making decisions that favor economics over lives, such may actually be the case.

Alexander Leaf, formerly head of the President's Scientific Commission on Aging, publicly stated, "To consider any extension of the human life span without a serious effort to anticipate and plan for the impact of increased longevity on society would be entirely irresponsible."

To sum up, the good news about chelation is that it probably will help you live longer and better. The bad news is that health insurers, the Social Security Administration, and Medicare probably can't afford to have that happen.

9

HARVARD SNUBS CHELATION

Mr. K., a married fifty-eight-year-old factory manager, first discovered he was a candidate for vascular surgery when he checked in for a series of tests after having suffered leg and chest pains.

Since Mr. and Mrs. K. are New York City residents, with easy access to the finest medical care and the most highly trained specialists, they were determined not to accept the first opinion, even coming from a highly credentialed cardiologist associated with one of New York's leading university-affiliated hospitals.

Mrs. K., well read and inquisitive, embarked on an intensive, month-long search of the medical and popular literature relevant to her husband's disease—and discovered chelation.

"Let's ask the doctor," Mr. K. said, never too quick to share his wife's enthusiasms. They did.

His response: "I don't know anything about it, except that it's no good and dangerous."

A scientific evaluation? Hardly.

An extraordinary response? All too common.

Although few well-trained professionals would be so brazen as to admit in the same breath that they are commenting on something about which they have little or no knowledge, the average physician who objects to chelation is equally ill informed, but just hides it better. Establishment medicine has long broadcast unfounded objections to chelation therapy, warning of documented dangers, for which there is little basis in fact.

To anyone contemplating chelation, such warnings arouse much distress. As a matter of fact, they caused the publisher of my earlier book, *Bypassing Bypass,* understandable concern. Just prior to contracting for this book, Mr. Sol Stein, president of Stein and Day Publishers, received the January 1983 issue of *The Harvard Medical School Health Letter*, denouncing chelation therapy for heart disease as an unproven medical treatment with clear evidence that it may be harmful.

With his permission, I will make you privy to our correspondence.

First, the communication the publisher received from Harvard Medical School, my alma mater:

> Chelation therapy for heart disease: Hundreds of physicians or clinics in the United States provide so-called chelation therapy in an attempt to open narrowed blood vessels in the heart. In the usual treatment program, a chemical known as EDTA is injected intravenously for one to two hours. The patient then rests for a few days, and the treatment is repeated for as many as 50 times. After weeks to months, at a cost of about $3,000, the patient is supposed to have better circulation to the heart, with a lowered risk of heart attack and fewer symptoms (such as pain or breathlessness).
>
> The rationale for this treatment is that EDTA binds calcium and removes it from the bloodstream. Because

calcium is found in the plaques that obstruct diseased arteries, proponents of chelation hope that lowering blood levels of calcium will allow calcium to dissolve out of the plaques, causing them to grow smaller. But, in fact, (1) the bulk of the material in a plaque is fiber, not calcium; (2) EDTA has not been proven to remove the calcium from plaques; (3) and, to date, no persuasive evidence from properly controlled studies has established that chelation therapy relieves symptoms.

An overdose of EDTA, rapidly administered, is potentially lethal. If given slowly, the drug is relatively harmless, although it can produce kidney damage. Temporary discomfort of one sort or another can also occur.

Chelation therapy does have established value as a treatment for heavy-metal poisoning, but it has not proven itself in heart disease. Anyone who accepts chelation therapy is automatically a "guinea pig," and unless the treatment is given as part of a carefully controlled trial, he or she is a guinea pig without purpose, as no useful information is going to be obtained.[1]

On January 6, 1983, the publisher wrote me the following letter:

I was concerned to see the January 1983 *Harvard Medical School Health Letter* refer to chelation therapy on its front page as an "unproven medical treatment"

---

[1]Excerpted from the January 1983 issue of *The Harvard Medical School Health Letter.*

I am much less concerned with their comments about an overdose being potentially lethal because an overdose of aspirin and a hell of a lot of other things are potentially lethal. I have some concern about the possibility of kidney damage but the major concern is with the fact that they say in the second paragraph that "to date, no persuasive evidence from properly controlled studies has established that chelation therapy relieves symptoms."

I'd welcome hearing your reaction at the earliest.

The following morning, I spoke with Mr. Stein by phone. Our conversation was cordial but to the point. He wanted to make certain that he would not be irresponsibly publishing material that might cause public harm. He made it clear that publication of this book hinged on my drafting a letter with supporting documentation to answer the questions raised in the *Harvard Medical School Health Letter*. He also informed me he intended to correspond with the editors of the *Health Letter* and was going to use my letter and my supporting documents to challenge their statements.

Fair enough. I had met a tough, but open-minded judge. It took me more than two weeks to draft my reply. On January 26 I wrote:

Dear Mr. Stein:

I am writing in answer to your questions concerning EDTA chelation therapy in the treatment of occlusive arterial disease. More specifically, I will point out what I consider to be errors in the January *1983 Harvard Medical School Health Letter (HMSHL)*. I am dismayed but not surprised that such a prestigious health letter, carrying an implied endorsement of the medical school from which I graduated, could be so misinformed about this safe and effective alternative to bypass surgery and to other treatments.

With the exception of occasional use in life-threatening cardiac irregularities caused by digitalis toxicity, or in the rare treatment of dangerously high blood calcium, I doubt whether contributors to the *HMSHL* have had sufficient personal experience to criticize EDTA chelation therapy. They seem unaware that it is the "disodium" form of EDTA which is used in the treatment of atherosclerosis, as distinguished from the "calcium-disodium" form used in the treatment of heavy metal poisoning.

There exists a difference of opinion between experts and, in this instance, I believe that I am far more expert in the use of EDTA chelation therapy than are contributors to the *HMSHL*. In addition to having been awarded my M.D. degree from Harvard Medical School, I am board certified as a specialist by both the American Board of Family Practice and the American Board of Chelation Therapy. I am presently Vice President of the American College for Advancement in Medicine, a professional association of physicians who practice and advocate chelation therapy. I have been using EDTA in the treatment of occlusive arterial disease for approximately ten years, with marked and lasting benefit in 75 to 95 percent of hundreds of patients treated. Observed benefits have correlated with the total number of treatments. No harm has come to any of those patients as a result of EDTA therapy.

The *HMSHL* states, "proponents of chelation hope that lowering blood levels of calcium will allow calcium to dissolve out of the plaques, causing them to grow smaller." Any such assertion is oversimplified and misrepresents the position of well-trained, chelating physicians. There is no evidence that chelation therapy reduces plaque size in humans. On the other hand, there is a wealth of evidence that the symptoms of reduced blood flow improve in more than 75 percent of patients treated.

Published research also indicates that treatment with EDTA improves the efficiency of energy metabolism independently of any effects on blood flow.

The *HMSHL* further states that EDTA is injected intravenously for "one or two hours." No physician qualified in the proper use of EDTA would infuse a therapeutic dose at such a rapid rate. Three hours is considered the minimum safe duration for an infusion. More rapid infusions have been responsible for kidney changes. Any medicine given in too high a dose over too short a time can cause harm.

The *HMSHL* correctly states that the therapy may last from weeks to months at a cost of about $3,000. They neglect, however, to point out that bypass surgery usually costs from seven to ten times that amount with a significant incidence of death, pain, suffering, and other serious complications, including long-term disability. Chelation therapy, properly administered, incurs no such risks. The cost of other non-surgical treatments can also equal or exceed the cost of chelation therapy over the same period of time. Cost figures must be considered in that perspective.

The *HMSHL* states that "no persuasive evidence from properly controlled studies has established that chelation therapy relieves symptoms." Physicians advocating chelation therapy are faced with a double standard. Bypass surgery was enthusiastically accepted without any such controlled studies. Coronary artery bypass has never been proven by "properly controlled studies" to be other than placebo effect. In fact, there is another operation called cardiac sympathectomy, which interrupts nerves that stimulate the heart with adrenalinlike substances. Cardiac sympathectomy also improves symptoms. It is impossible to perform bypass surgery without also interrupting sympathetic nerves, reducing spasm of coronary arteries and reducing workload of the heart. It has never been proven in "properly controlled studies" which effect is most important, bypass or sympathectomy.

Enclosed are nine recently published scientific papers concerning chelation therapy, many of which show statistically significant, objective measurements of improved blood flow following EDTA chelation. [Readers will find these studies described in greater detail in chapter 10 and elsewhere in this book, along with additional studies published after this letter was written.] The study by Dr. Casdorph on heart disease might be criticized because changes in cardiac ejection fraction were relatively small. Statistical analysis in that paper was based solely on the probability of measuring 17 improvements in 18 patients, irrespective of the degree of improvement. This is analogous to calculating the probability of flipping a coin 18 consecutive times and getting 17 heads. It is not necessary to consider the degree of change of cardiac ejection fraction to prove a high degree of statistical significance. Observed clinical improvement was also dramatic. Dr. Casdorph's second study, which measures brain blood flow, contains even more convincing data to prove that EDTA chelation therapy results in a very significant improvement in circulation to the brain. The probability that these results could have been due to random chance are less than five in ten thousand. That study was duplicated and confirmed independently by other researchers using a different technology.

A manuscript in press by Drs. Casdorph and Farr does not offer statistical proof of effectiveness, but that paper does describe four patients who had each been told to have a leg amputated. They sought out chelation instead and all four are now enjoying a good quality of life, able to walk on both legs without the recommended amputation.

Another study reports the results of kidney function tests done on 383 consecutive patients, before and after treatment with EDTA for chronic degenerative disease. That study showed no adverse effects from EDTA, and it did show a statistically significant improvement in kidney function following chelation. One patient, of the 383 studied, was at first an exception. She was an 86-year-old woman who began the treatments with preexisting abnormal kidney function and experienced further deterioration following EDTA. Three months following treatment her kidney function was closer to normal than when she began. Physicians qualified to administer intravenous EDTA give reduced dosages at less frequent intervals to patients with diminished kidney function.

One enclosure is a review of the scientific literature concerning kidney effects of EDTA. That review indicates that reports of kidney damage stem from patients with severe preexisting kidney disease prior to EDTA or who suffered with diseases frequently associated with primary kidney disorders. Also, during the early use of EDTA the dose-rate of administration was often six or more times higher than is now considered safe.

Temporary discomfort from EDTA mentioned in the HMSHL is quite minimal when compared with the discomfort of bypass surgery or even coronary artery angiograms.

The *HMSHL* reference to "guinea pigs" prompts a similar observation that anyone who accepts bypass surgery is also "automatically a guinea pig." Why is it that the critics of chelation should demand that millions of dollars be spent in large-scale, double-blind, controlled studies of chelation to prove that the benefits are not due to placebo effect, while at the same time they enthusiastically refer patients for bypass surgery for which no similar studies exist? The proponents of bypass surgery say that they do not need controlled studies to prove effectiveness because the benefits are so obvious. It is equally obvious to practitioners of chelation therapy that their patients also improve, for far less cost and at very little risk.

Critics of chelation therapy say that it is "unproven," but they ignore the fact that no funds have been made available for research with chelation. Committees that control government research funds seem to share a common bias with the critics. These same committees have allocated tens of millions of dollars toward research of bypass surgery while experts are still arguing about whether medical therapy might not be equally effective in many or most patients subjected to surgery.

The statistically significant studies that I have described might still be criticized because they are not double-blind. These studies were performed at the personal expense of physicians in practice, using patients as their own controls. That is, objective measurements were made on patients prior to therapy and follow-up measurements were made after therapy. Critics of chelation therapy may say that since both the patient and the treating physician knew that every patient was receiving EDTA, the observed benefits would be in the range of what one might expect from the placebo effect. With 75 to 90 percent of patients improving and with symptoms which had often not responded to all other available therapies, it is not likely that the placebo effect could be responsible for such universal benefit.

I consider the enclosed reports to be good science. All scientific observations begin with anecdotal reports and then with a small series of patients, prior to undertaking the very expensive, large-scale, double-blind studies in which half of the patients truly are "guinea pigs," and unknowingly do not receive the active ingredient under study.

Contributors to the *HMSHL* seem to be relying on outdated reports published ten to fifteen years ago for their opinions. I doubt that they are aware of the recently published data which I have cited. Human Sciences Press, publisher of the *Journal of Holistic Medicine*, in which six of the nine enclosures have appeared, informs me that the Francis A. Countway Library of Medicine at Harvard Medical School does receive the *Journal of Holistic Medicine* and that it should have been available to the *HMSHL* staff. In defense of the *HMSHL*, I might add that these recent papers would be somewhat difficult to find since the *Journal of Holistic Medicine* is not yet listed in the *Index Medicus*, nor are papers published in that journal available for a computer search on the topic of chelation therapy. The *Journal of Holistic Medicine* applied to the National Library of Medicine two years ago for listing in the *Index Medicus* and for availability to the MEDLINE computer search program. A committee which selects journals for such listings can approve only ten or fifteen percent of all publications received by the National Library of Medicine. Monetary and personnel limitations do not allow for a more comprehensive service. Many foreign language publications are not received by the National Library of Medicine and are therefore not even considered. A number of foreign research papers quite favorable to chelation therapy are in that category, and are listed as references in the enclosures. [These references are also listed at the end of chapter 17.] Most health care professionals are quite unaware that when they request a computer search on a specific topic they receive a representation of less than ten percent of what has been published in the world's scientific literature.

Experts with extensive experience in the use of chelation therapy routinely observe dramatic improvement in the vast majority of their patients and they do not feel justified in withholding this treatment if symptoms are present. The economics and medical politics of the situation are such that proponents of other therapies, especially bypass surgery and expensive prescription drugs, find themselves in a much more powerful position than that of chelating physicians. Big money and huge industries are involved. As chelation therapy is increasingly brought to public awareness, criticism from these more politically powerful sectors becomes stronger. The recent *HMSHL* is just one such example.

Approximately the same number of patients has undergone chelation therapy as has undergone bypass surgery. Most such patients have accepted chelation therapy on the recommendation of friends or family members who have previously been helped by chelation, despite the absence of insurance reimbursement and without benefit of the widespread media publicity and "status" associated with bypass surgery. They often chose chelation against the advice of their personal physicians and cardiologists. Chelation therapy is made available by physicians courageous enough to resist peer pressure and to provide a nontraditional treatment that they feel is safer and more effective.

To be completely fair, all forms of therapy should eventually be researched on equal terms. There is an urgent need for grant funding from the National Institutes of Health (NIH) to allow proper studies to further prove or disprove safety and effectiveness of all therapies. EDTA is not patentable and a pharmaceutical company could not recover the research costs of such a study. Practicing physicians cannot afford to do more than limited studies. That leaves only NIH as a source of funding for an "orphan" type drug; that is, a drug which has been around for so long that the patent has expired and for which an important new use is discovered.

I sincerely doubt whether the best interests of the American public are being served by existing health care procedures and policies.

Sincerely,

Elmer M. Cranton, M.D.

The most immediate result of this interchange of letters was the contract for the first edition of this book (since then, extensively revised and updated). But soon thereafter came a different, rather unexpected response from another quarter. I received a letter from an influential Harvard alumnus who works closely with the Harvard Medical School dean in fund raising. He wrote to tell me of his particular personal interest in chelation and in my rebuttal of the negative report in the *HMSHL*. He mentioned that he had had a quadruple arterial bypass three and one half years previously, and was good friends with a Texas physician and former classmate who happens to be one of the pioneers of chelation therapy in the United States.

Thinking he might one day benefit from chelation, he had communicated with the dean at Harvard to request they send him all the scientific material—research and test results—which had served as a source for the *HMSHL* chelation report.

"Only very recently," he wrote to me, "did I get the material requested. It was a copy of a news article that had appeared in the *Internal Medicine News* of September 14, written by a staff reporter, Mr. W. R. Kubetin. The article merely quotes the opinions of doctors Harrison and Frommer and others. It is not a scientific paper of course but a popular rendition of opinions of others who may or may not have been involved with the chelation procedure. I was somewhat disappointed that the *Harvard Letter* would not investigate primary medical and scientific sources before printing its report."

One thing he might have mentioned, but did not, was the biased way in which the *HMSHL* staff chose to quote sources from the news tabloid. They carefully included all the antichelation comments and just as carefully refrained from including any of the prochelation opinions or research.

This will not, of course, put an end to groundless antichelation charges, but it should serve notice to biased critics that they had better start doing their homework.

10

CLINICAL RESEARCH: ALL GOOD!

The medical community eagerly accepts scientific research buttressing a therapy it already approves. Somewhat more reluctantly, it examines and debates entirely novel approaches. But what it hates worse than poison is reappraising a treatment once rejected. Medicine, after all, is made up of people—people trailing M.D.s after their names—who, like the rest of us, do not enjoy admitting error.

Someday when chelation therapy is an established part of standard medical care, historians of twentieth-century medicine will wonder how so much supportive research on its benefits could have been scrupulously conducted by skillful medical researchers and even more scrupulously ignored by the guardians of our health. By that time, most of the individuals who successfully shifted chelation toward the fringes will not be alive to blush, sparing them extensive embarrassment.

The amount of positive research is certainly formidable. And those studies that purport to demonstrate that chelation doesn't work actually show the opposite. We will now examine much of this research in detail.

In a sense, we're attempting to set the record straight and

to tell people who read this book—especially physicians—where they should look for the scientific evidence. After all, mainstream medical journals engage in unconscionable editorial censorship. They refuse to publish positive research studies on EDTA chelation but are quick to print editorial criticism and anecdotal letters to the editor that are biased against this marvelous therapy. They are also quick to uncritically print highly flawed studies that erroneously allege to disprove chelation, as demonstrated by the Danish and New Zealand studies analyzed below. Journals that do publish supportive studies, although medically excellent, tend to be smaller, less widely read, and ignored by the mainstream. Studies supportive of EDTA chelation therapy have consistently been refused inclusion in the MEDLINE computer database by the National Library of Medicine.

Also, academically positioned researchers and professional clinical trialists have been chastised repeatedly by their colleagues, should they be intellectually honest enough to express an interest in research of EDTA chelation therapy for atherosclerosis. They are told behind closed doors that this is not a "politically correct" topic and that such a research interest would be "career suicide."

Most practicing physicians are entirely unaware that less than 20 percent of the world's total biomedical literature (in all languages) is referenced by the National Library of Medicine in the *Index Medicus* and its electronic counterpart, the MEDLINE computer database. Thus, a computer search for positive studies of chelation therapy in the treatment of atherosclerosis will be deceptively negative.

In this chapter, we will discuss several of the most important positive studies, referenced for those who wish to obtain the original articles. Then we will analyze the allegedly "negative" studies. A very complete listing of all studies thus far published on this topic can be found at the end of chapter 17.

Let me make a few points before we begin.

- First, there are no genuinely negative studies. All the medical research on chelation has produced positive results. That is, the data have invariably been positive. This has not prevented medical spinmeisters from occasionally imposing a negative interpretation on a positive result.

- Second, financial considerations have limited the size of chelation studies. The drug is no longer patentable and no one has been willing to spend the $30 million or more that pharmaceutical companies must spend to satisfy the FDA requirements before marketing claims can be made. That's the price tag for a large double-blind, placebo-controlled study. Many of the studies we quote are relatively small. Though they often have less than a hundred patients, they are nonetheless scientifically significant. Their endpoints are determined by objective numerical measurement of increases in blood flow and are statistically analyzed. In other words, conclusions were not determined by merely asking patients how they subjectively felt.

- Third, critics of chelation have frequently suggested that reported improvements are a placebo effect. It is a well-known phenomenon of medicine that when given a completely inactive substance and told that it may help them, many people will show a certain—often impressive—level of improvement for some time after they begin using their new "medicine." Thus, the placebo (as inactive substances used in medical research are called) turns out to have an "effect." But chelation hardly fits the profile of a placebo. A placebo effect begins shortly after its first administration and rarely, if ever, persists for more than six months. Chelation, by contrast, shows its full range of benefits quite slowly. Usually, it requires not only several months of therapy but also an additional several months after a course of therapy for the full benefit of treatment to occur. And the benefits generally persist for years. Therefore, chelation shows a pattern different from, and indeed opposite to, the pattern of a placebo. Moreover, when studied properly, the benefits are far stronger than a placebo could show. It's nonsense to allege that such dramatic improvements are placebo effects.

• Fourth, statistical analyses of measured improvements in the more carefully performed chelation studies demonstrate that the probability of these changes being due only to random chance is somewhere near the vanishing point of statistical insignificance. That numerical probability ranges from less than one in one thousand to less than one in ten thousand. The reason for such high significance is the magnitude of the improvements measured, despite the relatively small number of patients.

With those general points to guide us, it's time to look at some of the actual studies done on atherosclerosis and chelation.

If you will recall the clinical trials I discussed briefly in chapter 1, they all clearly demonstrated improved circulation after chelation. These are the sorts of results that any chelation therapist expects—we not only notice improved exercise tolerance, memory, and mental alertness in our patients but even healthy color returning to their cheeks.

Many other objectively measured indicators of circulatory health tell similar stories. Drs. McDonagh, Rudolph, and Cheraskin took seventy-seven elderly patients with documented narrowing of the peripheral arteries in their legs and measured changes in blood flow after approximately twenty-six EDTA infusions administered over sixty days. They used the preferred method for such testing: the ankle/brachial Doppler blood pressure ratio. This method compares the blood pressure in the arms with that in the ankles using Doppler ultrasound. In a person with a youthful circulatory system, the normal pressure in the ankles is equal to or greater than that in the arms. Patients with impaired circulation to the lower extremities have, of course, weaker arterial blood flow and lower blood pressure in their ankles than in their arms.

On average, after chelation therapy, the patients' ankle pressure increased from 55 percent of the arm pressure to 71 percent of the arm pressure, a change so significant that the statistical likelihood of its being due to random chance would be somewhere in the neighborhood of one in ten thousand.[1] Improvement in Doppler blood pressure reflects only blood flow in larger arteries. EDTA also improves capillary circulation, which is especially reduced in diabetes.

Drs. Casdorph and Farr reported on four patients who had all been recommended to undergo surgical amputation of their gangrenous lower extremities before treatment with EDTA chelation. Clearly, these were people who had reached end-stage complications of atherosclerosis and poor blood flow. Most of them had deep ulcerations and large areas of dead, necrotic tissue on their feet. In some cases, circulation to the extremities had become so poor and so much tissue had died that the condition was no longer causing significant physical pain. The patients' pain was now mental—in the clear knowledge that they were about to lose a leg.

All four patients chose to postpone amputation (against surgical advice) and receive infusions of EDTA. Treatment was completely successful in three out of the four cases.[2] In the fourth case, the patient did eventually lose the tips of his second, third, and fourth toes, but the foot and leg were saved. After chelation, all four patients recovered circulation in their lower extremities sufficient to not only protect them from amputation but to also allow them pain-free walking without limitation or handicap. Several years after chelation therapy, those four patients continued to be alive and well, walking on their own legs and feet. Their recovery—if witnessed by a physician who was unaware of or unwilling to credit chelation's effectiveness—could only be seen as a sort of medical miracle, something comparable to spontaneous remission of an advanced and deadly cancer.

Another study by Dr. Casdorph, as mentioned in chapter 1, contains data showing large, numerically tabulated and objective improvements in blood flow to the brain. Computerized graphs showing improved blood flow are astounding, even to those untrained in medical science. A scintillation counter and computer were used to generate those sophisticated images, which are perhaps the most convincing objective evidence we have for increased blood flow after chelation. I challenge any open-minded physician to review the data in that article and not come away impressed.[3]

Serious students of chelation therapy and health care professionals are referred to a *Textbook on EDTA Chelation Therapy, Second Edition*, edited by Elmer M. Cranton, M.D., which is now available from Hampton Roads Publishing Company, Inc., 1125 Stoney Ridge Road, Charlottesville, Virginia 22902, (434) 296-2772, FAX (434) 296-5096. Complete copies of the chelation research studies cited in this chapter, including the actual data, are republished in that book.

In addition to those smaller studies, there have been other large retrospective studies using a variety of methods to measure changes following chelation therapy.

Drs. Hancke and Flytlie, two Danish doctors with impeccable credentials, published such a study in 1993—a counterblast, as it were, to the Danish bypass surgeons' ineffectual attempt to discredit chelation in the previous year, as described later in this chapter. Hancke and Flytlie measured improvements using several different criteria in a series of 470 patients who were followed for six years following chelation therapy. Of 265 patients with coronary artery disease and narrowing of the blood vessels to the heart, they reported improvement in 90 percent. Sixty-five of those patients had been referred for bypass surgery before chelation. After treatment, 58 of the bypass candidates improved so dramatically that they avoided the surgeon's knife. Among the 207 angina patients using nitroglycerin to control their pain, 189 were able to reduce their consumption. Most discontinued its use altogether. Of 27 patients awaiting foot or leg amputation, 24 avoided surgery.[4]

These results, remarkable as they may seem, fully correspond with what physicians who administer chelation therapy routinely observe in practice.

Another even larger retrospective study done in Brazil, analyzed the effects of chelation on 2,870 patients with atherosclerosis and related degenerative conditions. Treatment was carried out between 1983 and 1986. Nearly all patients were being treated for atherosclerotic vascular diseases. The most serious category, the patients with heart disease, numbered almost one-third of the total; and, in that group following chelation, 77 percent showed marked improvement, 17 percent showed good improvement, 4 percent had partial improvement, and 3 percent were unchanged or worse. Patients with arterial blockage in other parts of the body showed similar improvements.[5]

The researchers, Dr. James Carter, a professor at Tulane University Medical School in New Orleans, together with Dr. Efrain Olszewer in Sao Paulo, Brazil, decided to follow up these treatment results by conducting a small double-blind pilot study on ten patients. Midway through the study they were forced to open up the blind for ethical reasons; five of the patients—these turned out to be the ones receiving EDTA chelation—were doing dramatically better than the placebo group. The placebo patients were then put on EDTA and they, too, rapidly began to improve.[6]

One final question is worth asking: Are these diverse studies (impressive though they may be) really typical medical research results on chelation? Drs. Terry Chappell and John Stahl set out to answer that question in 1993. They conducted a metanalysis of all currently available scientific literature. This is an eagle-eyed observation and comparison of diverse studies that summarizes, as best it may, the total results achieved by many researchers following chelation therapy. Over the course of the last decade, such analytic overviews have grown more reliable and are more relied upon. Chappell and Stahl identified nineteen articles in the medical literature that met their criteria for determining chelation's effectiveness in cardiovascular illness. In combination, the articles provided data on 22,765 patients. The metanalysis determined that 87 percent of these patients experienced favorable outcomes. Only those improvements measured by objective testing were accepted as evidence in their analysis.

Chappell and Stahl were compelled to conclude that there was very strong published evidence for chelation's effectiveness in the treatment of cardiovascular disease.[7]

There is nothing surprising about such a conclusion. It's very difficult to test real people using chelation therapy and not come away impressed. Nevertheless, some physicians have achieved that feat. Let's look at their research.

Every now and then puzzled patients tell me that a friend, relative, or skeptical physician has told them that chelation was fairly tested and fell flat. I can usually guess what they're referring to. In the last ten years, a small cluster of studies sprouted up in the medical literature purporting to demonstrate that EDTA chelation was a fizzle when it came to treating cardiovascular ailments.

The curious thing is that those studies—flawed and imperfect though they are—only succeed in offering us still more positive data to support this therapy.

The most controversial and oft cited study was done in Denmark. It was the handiwork of a group of Danish cardiovascular bypass surgeons. Results of that study were published in two medical journals, the *Journal of Internal Medicine* and the *American Journal of Surgery*. The results were also widely publicized in the news media.

The surgeons had taken 153 patients suffering with intermittent claudication. These were people with such severely compromised circulation in their lower extremities that walking across a parking lot could challenge their fortitude. One measurement of their condition was their maximal walking distance (MWD)–the very longest distance that they could walk before intolerable leg pain brought them abruptly to a halt. The patients were divided into an EDTA group and a placebo group. In the pretreatment phase, the EDTA group could on average walk 119 meters before colliding with their MWD; the placebo group averaged 157 meters.

Treatment began with the patients receiving either twenty intravenous infusions of EDTA or twenty infusions of a simple salt solution, depending on their group. The study was purportedly double-blinded, that is, neither the patients nor the researchers knew which person was receiving which infusion until after the study was complete. Progress was measured periodically. In particular, we will analyze their results at three months following treatment, when full benefit from chelation would be expected to occur.

Both placebo and treatment groups showed improvement. However, the investigators concluded that the improvement was not statistically significant and—equally important—that the difference in response rate between the EDTA group and the placebo group was roughly similar. Obviously, a drug that fails to achieve more than the placebo effect is presumed to be a dud.

The Danish study impressed many people; but, in rather short order, the integrity of the study was called into question. It was learned that the researchers had violated their own double-blind protocol. Not only did they themselves know before the end of the study who was receiving EDTA and who placebo, they had also revealed this information to many of the test patients. Before the study was over, the researchers and more than 64 percent of the patients were aware of which treatment they had received.

This method was unorthodox and, since it had not been reported in the published study, extremely questionable from an ethical standpoint.

Many people had also been struck by the study's relatively small size. Intermittent claudication is a very unpredictable disease, and, unless enough patients are included in a trial, the results tend to be statistically unreliable.

The most interesting aspect of the Danish study, however, was hidden away in the numbers. This is the startling fact that the patients who were given EDTA were certainly a good deal sicker than the patients tested with a placebo. Therefore, the improvements they made were harder earned and more significant. The researchers, who candidly admitted that they undertook the study to convince the Danish government not to pay for chelation, either never noticed that aspect or felt reluctant to reveal it. The evidence is in the pretreatment MWDs. The EDTA patients' longest average distance before claudication pain stopped them in their tracks was 119 meters, while for the placebo patients it was 157 meters.

Still more significant was the standard deviation. Standard deviation is a statistical abstraction, which reflects the amount of variability among a collection of raw scores. In essence, standard deviation reflects how widely diverse the numbers are in each group. A high standard deviation indicates that measurements were spread out toward the extremes of a wide range, rather than closely clustered near the average. Without going further into the arcane science of statistics, it is enough to say that the plus or minus 38 meters for EDTA patients versus plus or minus 266 meters for the placebo group represents an enormous difference in walking capacity that is heavily biased in favor of the placebo group. The standard deviation numbers show that some placebo patients must have walked half a mile before stopping. The EDTA group's claudication was therefore much more severe. The EDTA group was much sicker. The design of the study was therefore catastrophically biased against EDTA chelation from the outset.

Yet, when the six-month study was completed, the MWD in the EDTA group rose by 51 percent, from 119 to 180 meters, while the mean MWD in the placebo group rose only 24 percent, from 157 to 194 meters. In plain English, looking at all the published data, the chelation group's improvement was more than twice as great as the placebo group's, even though they were significantly sicker at the outset.[8,9]

I believe the Danish study must be interpreted as another solid demonstration of the effectiveness of chelation. If it were not for its relative smallness, I would be happy to quote from its results at any time. I hope the Danish surgeons can be persuaded to undertake another study with five times as many subjects. If they take the trouble to hire an academic statistician to oversee design and interpretation of the study, and refrain from violating the double blind, they may yet do good work, and we shall all be much in their debt.

Another study—also conducted by vascular surgeons—was done at the Otago Medical School in Dunedin, New Zealand, two years later. The subjects of this study were also suffering from intermittent claudication manifested by leg pain and walking difficulties beyond a very limited distance. Chelation subjects were compared to controls. The study extended to three months after twenty infusions of either EDTA or a placebo had been administered. Upon examining the results, the authors of the study concluded that chelation had been ineffective. Once again, that conclusion seems ill founded.

The absolute walking distance of the EDTA group increased by 26 percent; in the placebo group, it increased by 15 percent. This was not considered statistically significant. The study, however, was so small that there were only seventeen subjects in the placebo group. One of these was what the statisticians call an "outlier." That is a person whose results differ strikingly from everyone else in the group. That placebo patient's walking distance increased by almost five hundred meters. All of the statistical gain in the placebo group was due to this one individual's progress. Without him, their placebo distance decreased slightly.

This illustrates the perils of a small study. A 25 percent gain in the EDTA group compared to no gain in the placebo group would have been very significant statistically.

Meanwhile, even the New Zealand researchers conceded that the improvement in artery pulsatility (measurement of pulse intensity) in the EDTA group's worse leg reached statistical significance. In statistical terms, there was less than a one in a thousand chance that that improvement was not a benefit of EDTA.[10]

I would note only two other things. First, a 26 percent improvement in walking is by no means minor and would attract notice if the agent had been a patentable drug. Second, even that level of improvement is in no sense representative of the much greater improvements claudication patients normally experience after chelation.

There is a simple reason for the difference: smoking.

Smoking so dramatically undermines cardiovascular function, especially in people who are already seriously sick with claudication, that it negates much of the gain that chelation provides. In the New Zealand study, 86 percent of the chelated subjects were smokers. They were advised to quit smoking when the study began, but how many of them actually stopped is, I fear, a subject for skeptical speculation. A demonstration of chelation's full potential requires a much higher percentage of non-smoking subjects at the outset.

Just as this book goes to the printer, another small study alleging to disprove EDTA chelation therapy is being widely reported by the news media. This recent study was conducted by cardiologists in Calgary, Canada, who freely admit their bias against chelation. They seem to have set out to discredit a therapy that they oppose by studying a few patients with heart disease. Because the study has not yet been published in a scientific journal, it is not possible to provide a meaningful critique. I feel certain, however, that when we finally do have an opportunity to conduct a detailed review of that study's design and data, the final assessment will be very similar to that of the Danish and New Zealand studies described previously—another hatchet job (see addendum on page 366).

It's relatively easy to design a study specifically to discredit an unpopular therapy, and to make that study superficially appear to be scientific. The United States Congress once commissioned its Office of Technological Assessment to analyze all published medical research for scientific merit. After a careful review of research studies from leading medical journals, they concluded that, "more than 75 percent of all published medical research has invalid or insupportable conclusions as a result of statistical problems alone." The final report to Congress stated, "few published clinical trials are well enough designed to yield valuable results."

And it's not merely intellectual dishonesty. Many doctors

who oppose chelation therapy firmly believe that it is ineffective. That is what they have been told. So they attack, with no personal knowledge about what they are attacking. Perhaps they feel threatened because very few doctors have the time to thoroughly read and analyze published studies in medical journals. They usually skim the abstract and jump to the authors' conclusions, accepting them without question.

I have also found medical doctors to be naive and unaware that the peer review process is often used as a form of editorial censorship—a way to maintain the status quo and protect the professional reputations and practices of the reviewers. Also, because medical journals so often depend heavily on advertising by major pharmaceutical companies, studies that are unpopular with that industry are rarely published, while brief letters to the editor and unsupported editorial opinion attacking such opposed therapies quickly find their way into print. Journals tend to be reluctant to bite the hand that feeds them.

Powerful psychological defense mechanisms also come into play. If doctors are not taught about EDTA chelation therapy in medical school (and they are not), and if those doctors therefore do not routinely use or prescribe chelation therapy for patients, then they believe one of two things: 1) either their medical educations were deficient and they are not providing the best of care for patients; or 2) other doctors routinely using and prescribing chelation therapy for medical conditions that are not FDA-approved must be "quacks," exploiting desperate patients. Which do you think their choice will be? It's apparently difficult for many medical doctors to shed an attitude of God-like omniscience and admit that they simply do not know everything there is to know.

One final study that was carried out with what I am forced to call negative intent is such a curious oddity that it also deserves discussion, although it remains unpublished. It is usually referred to as the "Heidelberg Trial" and was conducted at the behest of the German pharmaceutical company Thiemann, AG, in the early 1980s. Once again using patients with intermittent claudication, it compared the effects of twenty infusions of EDTA with twenty infusions of bencyclan, a vasodilating and antiplatelet agent owned by Thiemann.

Needless to say, from a practical commercial standpoint, Thiemann's action was bizarre. If EDTA did well in the trial, Thiemann's own already well established drug could only suffer. Nonetheless, the trial went forward and was reported before the audience in 1985 at the Seventh International Congress on Arteriosclerosis in Melbourne, Australia. That study showed that immediately following twenty infusions of EDTA, pain-free walking distance increased by 70 percent. By contrast, the patients receiving bencyclan had increased their pain-free walking distance by 76 percent. The difference between these two results was, of course, not statistically significant, but another result was. It turned out that twelve weeks after the series of infusions was completed, the EDTA patients' average pain-free walking distance had continued to increase, going up by an astounding 182 percent. No further improvement had occurred in the patients receiving bencyclan, however.[11]

A report from Thiemann only mentioned the 70 and 76 percent figures, and press releases stated that chelation was no better than a placebo without mentioning that the "placebo" was a drug that had been proven effective in the treatment of intermittent claudication. Thiemann never released the actual data from the Heidelberg Trial, but some German scientists who had access to it, and who were disturbed at the deception they were witnessing, chose to reveal the data to members of the American scientific community.

The complete data showed that four patients in the EDTA group experienced more than a one-thousand-meter increase in their pain-free walking distance following treatment.[12] This highly favorable data from those four patients mysteriously disappeared before the final results were made public. By sponsoring and funding the study, Thiemann had a legal right under terms of their contract to edit the final results and to interpret the data in any way that suited them. An analysis of the complete data showed an average increase in walking distance in the EDTA-treated group of 400 percent at three months after therapy—five times the 76 percent increase of the group receiving bencyclan.

These three ineffectual attempts to discredit chelation with flawed research represent pretty much the sum total of scientific involvement that the establishment has had with this extraordinary therapy over the past thirty years.

However, the darkest moment for chelation actually came way back in 1963. This was when Drs. J. R. Kitchell and L. E. Meltzer coauthored an article reassessing their support for EDTA chelation.

Although it was hardly in widespread use, chelation had been surprisingly uncontroversial up until that moment. Beginning in 1953 Dr. Norman Clarke and his associates at Providence Hospital in Detroit began using EDTA chelation to treat coronary heart disease. In 1956 they reported that they had treated twenty patients suffering from chest pain (angina pectoris). Nineteen of the twenty patients had had a "remarkable" improvement in symptoms.[13]

Soon other physicians became interested, among them Drs. Kitchell and Meltzer, who specialized in cardiology at Presbyterian Hospital in Philadelphia. From 1959 to 1963, Kitchell and Meltzer reported on their consistent good results treating cardiovascular diseases with EDTA. Their early reports were all very positive.[14,15,16]

But in April of 1963, shortly after their last favorable report, they published a "reappraisal" in the *American Journal of Cardiology* that questioned chelation's value.

That reappraisal article included ten original patients on whom they had previously published data and twenty-eight patients with coronary heart disease who were treated subsequently. Treated patients in this report were all severely ill. The authors state that the patients were "referred to us because of severe angina. The patients had previously been treated with most of the accepted methods, and their inclusion in this study resulted from wholly unsuccessful courses. Each of the patients was considered disabled at the start of therapy." This was therefore a very high-risk group with any form of therapy.

Seventy-one percent of patients treated had subjective improvement of symptoms, 64 percent had objective improvement of measured exercise tolerance three months after receiving twenty chelation treatments, and 46 percent showed improved electrocardiographic patterns. Kitchell and Meltzer then went on to conclude that chelation was not effective because some patients eventually regressed more than a year after treatment. However, considering the poor health of the patients, there is no other treatment about which the same statement could not be made. Eighteen months following therapy, 46 percent of those patients remained improved. The results were very favorable even though the authors' conclusions were not.[17]

I believe that this "reappraisal" article was largely responsible for termination of academic research into chelation as a treatment for cardiovascular ills. Rather than analyzing the data for themselves, most physicians simply accepted the mistakenly negative conclusion at face value. We will probably never know what prompted those early researchers to change their position so abruptly. We can only speculate that it was an unrealistic expectation that the emergence of bypass surgery would be a final solution.

The years that followed were filled with astonishing demonstrations of surgical inventiveness; and, for at least the next two decades, cure by the knife dominated the medical landscape. Then came balloon angioplasty. Those surgical and high-tech discoveries were splendid in themselves; but what was tragic was to regard them as the preferred, if not the exclusive, approach to complex cardiovascular problems.

As for chelation, its future is now bright because its effectiveness is incontrovertible. Biased or uninformed physicians may call it untested, but no scientifically informed person can read the studies on which this chapter is based without realizing that EDTA chelation therapy is a formidable antagonist to cardiovascular disease.

A major upsurge in demand for chelation is now coming from the many people who have heard firsthand from friends or relatives who benefited from this remarkable therapy.

Periodic research updates will be posted on my website: www.drcranton.com.

References

1. McDonagh EW, Rudolph DO, Cheraskin E. "Effect of Chelation Therapy plus Multivitamin/Mineral Supplementation upon Vascular Dynamics: Ankle/Brachial Doppler Blood Pressure Ratio," *Journal of Advancement in Medicine* 2, nos. 1&2 (1989): 159-66.

2. Casodorph HR, Farr CH. "IRDTA Chelation Therapy: Treatment of Peripheral Arterial Occlusion, an Alternative to Amputation," *Journal of Advancement in Medicine* 2, nos. 1&2 (1989): 167-82.

3. Casdorph HR. "EDTA Chelation Therapies II, Efficacy in Brain Disorders," *Journal of Advancement in Medicine* 2, nos. 1&2 (1989): 131-54.

4. Hancke C, Flytlie K. "Benefits of EDTA Chelation Therapy on Atherosclerosis: A Retrospective Study of 470 Patients," *Journal of Advancement in Medicine* 6, no. 3 (1993): 161-71.

5. Olszewer Z, Carter JP. "EDTA Chelation Therapy: A

Retrospective Study of 2,870 Patients," *Journal of Advancement in Medicine* 2, nos. 1&2 (1989): 197-213.

6. Olszewer E, Sabbag FC, Carter JP. "A Pilot DoubleBlind Study of Sodium-Magnesium EDTA in Peripheral Vascular Disease," *J Natl Med Assn* 82, no. 3 (1990): 174-77.

7. Chappell LT, Stahl JP. "The Correlation between EDTA Chelation Therapy and Improvement in Cardiovascular Function: A Metanalysis," *Journal of Advancement in Medicine* 6, no. 3 (1993): 139-60.

8. Guldager B, Jelnes R, Jorgensen SJ, et al. "EDTA Treatment of Intermittent Claudication—A Double-Blind, Placebo-Controlled Study," *Journal of Internal Medicine* 231 (1992): 261-67.

9. Sloth-Nielsen J, Guldager B, Mouritzen C, et al. "Arteriographic Findings in EDTA Chelation Therapy on Peripheral Atherosclerosis," *The American Journal of Surgery* 162 (1991): 122-25.

10. Van Rij AM, Solomon C, Packer SG, et al. "Chelation Therapy for Intermittent Claudication: A Double-Blind, Randomized, Controlled Study," *Circulation* 90, no. 3 (September 1994): 1194-99.

11. Diehm C, Wilhelm C, Poeschl J, at al. "Effects of EDTA Chelation Therapy in Patients with Peripheral Vascular Disease—A Double Blind Study," an unpublished study performed by the Department of Internal Medicine, University of Heidelberg, Germany, in 1985. Presented as a paper before the International Symposium of Atherosclerosis, Melbourne, Australia, October 14, 1985, Podium Presentation.

12. Carter JP. "The First Double-Blind Chelation Study in the Treatment of Vascular Disease," University of Heidelberg Medical School, 1985; a transcribed interview with researchers involved in that study.

13. Clarke NE, Clarke CN, Mosher RE. "Treatment of Angina Pectoris with Disodium Ethylene Diamine Tetraacetic Acid," *Am J Med Sci* 232 (1956): 654-66.

14. Kitchell JR, Meltzer LE, Seven NJ. "Potential Uses of Chelation Methods in the Treatment of Cardiovascular Diseases," *Prog Cardiovasc Dis* 3 (1961): 338-49.
15. Kitchell JR. "Peripheral Flow Opened Up," *Medical World News* 4 (March 15, 1963): 36-39.
16. Meltzer LE, Ural ME, Kitchell JR. "The Treatment of Coronary Artery Disease with Disodium EDTA," in Seven MJ and Johnson LA, eds., *Metal Binding in Medicine* (Philadelphia: J. B. Lippincott Co., 1960), 132-36.
17. Kitchell JR, Palmon F, Aytan N, et al. "The Treatment of Coronary Artery Disease with Disodium EDTA: A Reappraisal," *American Journal of Cardiology* 11 (1963): 501-6.

11

THE REAL DANGERS YOU HAVEN'T BEEN WARNED ABOUT

Heard from your doctor lately? Has he called to check up on how you and your family are doing, or let you in on the latest news from his medical society? Does he keep tabs on you to make sure you don't inadvertently stumble into medical trouble? No?

If you live in Lincolnton, North Carolina, you might get a very special type of personal medical service. You don't need to subscribe to the *Harvard Medical School Health Letter* to be provided with misinformation about the alleged dangers of chelation therapy either. All you need do is make an appointment to visit a chelating physician.

At least that was Bobby C.'s experience. Two days after he called my office and scheduled a visit, he received the following *unsolicited* communication (a copied document from the Council of Scientific Affairs of the American Medical Association) from John R. Gamble Jr., M.D., a Lincolnton physician. It read:

SUBJECT: Chelation Therapy
QUESTION: How safe and effective is chelation therapy

with ethylenediamine tetraacetic acid or its sodium salt for the treatment of atherosclerotic vascular disease? ANSWER: There was a consensus among all respondents that chelation therapy with ethylenediamine tetraacetic acid or its sodium salt was not an established treatment for atherosclerotic vascular disease.

The original thesis that repeated intravenous infusions of the chelating agent, EDTA disodium, was of benefit to patients with coronary artery disease, as manifested by the anginal syndrome, has not been established in any well-designed, controlled trial. Although some uncontrolled studies claim positive benefits, others have shown no significant effects from such therapy. There is no supporting evidence that it has any significant effect upon the atherosclerotic plaque.

Furthermore, the safety of using EDTA, especially in patients with coronary artery disease is questionable. Chelation of plasma calcium will decrease the levels of ionized calcium and result in tetany, cardiac arrhythmias, convulsions, and respiratory arrest. It can cause renal tubular necrosis and renal failure, permanent renal damage, bone marrow depression, and prolongation of the prothrombin time.

The majority of the respondents felt that this treatment was unacceptable or indeterminate therapy for atherosclerotic vascular disease. About half as many felt that it could still be considered investigational, i.e., worthy of a controlled trial under protocol.

The above response is provided as a service of the American Medical Association. It is based on current scientific and clinical information and does not represent endorsement of the AMA of particular diagnostic and therapeutic procedures or treatment.

And what did this prestigious organization, the AMA, cite as its reference? You guessed it. Outdated and incomplete reviews, composed primarily of editorial opinion by physicians with little or no chelation experience.

In Corpus Christi, Texas, a group of physicians, equally eager to warn the public of potential danger, banded together in a mass media approach. The following advertisement, without mention of a sponsoring organization, but with an alphabetical listing of 247 doctors, appeared in the local daily, the *Corpus Caller:* "MEDICAL NOTICE—The undersigned doctors agree that there is no good scientific evidence for any benefit from Chelation Therapy for heart disease or hardening of the arteries. Furthermore, it is probably of no value in that type of illness. It is worthy for the public to note that this method of treatment has not been approved by the FDA."

What was going on in Corpus Christi?

Curious, we began calling physicians listed, in alphabetical order. Of the first eleven contacted, only two were available to discuss their antichelation position.

Both referred us to Jean Oliver, at the Nueces County Medical Society, explaining they knew very little about it (the ad or chelation), but stated that "it" (chelation) was of "no value" and "dangerous."

Although each of these physicians was noticeably reluctant to speak to the issue, the first said he felt the "danger would be to make the patient believe that he is getting cured of something that has not been proved." The second doctor explained that EDTA takes calcium "out of the bones and everywhere else," but "probably does not take it out of arteries . . . You could get osteoporosis, weakness in the bones, lose most of your strength out of your skeleton, and suffer bone pains."

When we called the offices of the Nueces County Medical Society, Mrs. Oliver denied the society had placed the ad and suggested we call Dr. Mario Eugenio, president of the group. Finally we reached Dr. Eugenio.

What was happening in Corpus Christi relative to chelation therapy?

"There have been a few problems."

"What kind of problems?"

"I am not in a position to tell you."

At the last, Dr. Eugenio offered the following: "There are some doctors using it, and charging $3,000, and even wives are complaining their husbands are spending the money that they should be giving in the house for a treatment that would not be acceptable. That's what started us looking around."

But if we really wanted to understand the "best positions" on the dangers involved with chelation, he suggested we go "right to the source" and contact the American Medical Association.

Although there is little doubt it is more dangerous to drive to the doctor's office than to be chelated once you get there, there are risks, and everyone considering chelation should be aware of them.

With more than forty-five years of documented clinical experience with EDTA to guide them, chelating physicians today have access to valuable information unavailable to their predecessors when it comes to taking every possible precaution to safeguard patients against the possibility of injury.

Even so, it is quite possible that chelation therapy has resulted in unpredictable harm in some isolated instances because of the unusual idiosyncrasies of individual patients. Without being facetious, there are citings in the medical literature of individuals so superallergic that they have died eating a peanut butter sandwich. The odds of such an occurrence are probably one in 50 million, but there are people who have had unfavorable reactions to almost any substance you can think of, peanuts included.

To the best of my knowledge and experience, it is extremely rare that a person is so allergic that he cannot be safely chelated, provided proper precautions are taken. When a patient does have an allergic reaction, it is rarely serious and can be immediately counteracted or neutralized by a physician who is well trained and competent. In my experience, in just about every such instance the sensitivity has not been to EDTA, but to one or more of the ingredients we add to the intravenous solution (vitamins and minerals, for example), which are sometimes manufactured with preservatives. When we detect a patient with allergies, we prepare a modified EDTA solution, custom-tailoring the ingredients to the individual sensitivity.

The EDTA solution includes some sodium, and there may be a potential hazard for patients with heart failure, whose cardiac status is extremely precarious. In those cases, where the infusion of a small amount of fluid and sodium (salt) might temporarily worsen a patient's condition, we carefully regulate diuretics, run the infusion more slowly, space the treatments further apart, and monitor all aspects carefully. With those precautions, most such patients can be chelated safely.

For many years it was thought that people with arrested tuberculosis should not be chelated because calcification of the tuberculoma is one factor believed to keep the disease in remission and prevent it from spreading. Now we know that EDTA's effect on that type of calcium is so minimal that there is only the remotest possibility of dissolving the calcium out of the calcium-coated granuloma in the lung wall. There have been no documented cases of this ever happening. But, just to be certain, we do periodic chest x-rays on patients with a history of arrested tuberculosis.

Another worry that has proved unjustified over time is that patients with calcium kidney stones might be adversely affected if the EDTA partially dissolved the stones, which might then be dislodged and get trapped in the lower urinary passages leading to the bladder. This is only a potential risk in someone who has a large stone upstream, where the interior of the kidney has a large open cavity. In such cases, of course, a kidney stone can become dislodged and cause a blockage to kidney drainage at any time, with or without chelation. Surgery may then be needed. Using high-powered ultrasound, so-called lithotripsy, most such stones can now be dissolved and passed without surgery.

Kidney stones that are made primarily from calcium may be dissolved by EDTA. There have been cases reported where, after fifty to one hundred or more infusions, they were not only partially dissolved but were then passed safely. In most instances, whether one has kidney stones or not has little bearing on the safety of being chelated.

I once treated a patient with diffuse kidney calcification stemming from polycystic kidney disease. This was causing progressive kidney failure. Following chelation he excreted much of this calcium as fine particles in his urine, and his kidney function greatly improved. He's convinced that chelation therapy has greatly delayed or prevented his need for dialysis or kidney transplant.

One more bugaboo that deserves burying is the worry that EDTA impacting on calcified arteries might unseat plaque and loosen chunks that would eventually block smaller downstream blood vessels, so-called plaque emboli. Plaque dislodgement is always a danger in patients with such atherosclerosis. But the fact is that it has never been a documented complication of chelation therapy when administered in currently accepted doses and rates—a remarkable history in light of the enormous number of patients treated over the years. Sudden onset of symptoms caused by that type of plaque embolism has often been what brought patients to me for chelation, but these symptoms have reduced in frequency or stopped altogether after a course of chelation therapy.

There is some small possibility of EDTA chelation resulting in temporarily diminished kidney function but only under very specific circumstances. If EDTA is administered too rapidly or in too large a dose for the patient's tolerance, the excreted toxic wastes can overload the kidneys, leading to impaired function. This is reversible if detected early and further chelation treatments delayed. If this overload is undetected, repeated chelations could cause prolonged kidney failure. The patient could then require renal dialysis. This has happened. Your best protection is to be absolutely certain the doctor you choose knows what he is doing. (More on this later.)

To put these possibilities in proper perspective, be reminded that there is no medicine known to man that cannot cause harm if too much is given too fast. Every physician knows that.

Digitalis-type medications, for instance, used routinely for heart conditions, have a long half-life, meaning they stay in a patient's system so long that it is easy to overdose. The difference between a therapeutic and toxic dose is so small that a slight overdose can turn out to be lethal. One of the most unforgettable things I was told at medical school was that there is hardly a doctor alive who uses digitalis-type medicines to any extent who hasn't killed a patient with an inadvertent overdose.

What is true of every other medicinal substance is also true of EDTA, except that it has very low toxicity when compared with most other medicines. It has a half-life of less than one hour in the body and is rapidly excreted unmetabolized, merely bound to metals. It has a remarkable safety record. After more than four decades of use in the United States and elsewhere, involving in excess of one million patients and more than twenty million treatments, it has triggered fewer untoward effects than aspirin.

What better evidence than the notable absence of malpractice claims against chelating physicians? A thorough search of available case law conducted by students at the Georgetown University Law School for the National Association of Insurance Commissioners while I was writing the first edition of this book failed to turn up any reported legal actions involving the use of chelation for atherosclerosis.

Considering the controversial nature of chelation and the large number of treatments that have been given, it is indeed noteworthy that chelating physicians have not been subjected to the large number of medical malpractice lawsuits that have plagued their more traditional colleagues.

There have probably been some kidney failures—even kidney deaths—with chelation therapy, but to the best of my knowledge this only happened when the doctors gave too much, too fast, to patients already at peril because of compromised kidney function. To condemn chelation therapy because there have been careless applications by underqualified physicians is tantamount to ruling out anesthesia for surgery because patients have suffered permanent brain damage or have died when an absent-minded anesthesiologist turned the wrong dial.

Try as they might, chelation critics have been unable to find more than a handful of patients harmed over the entire four and one-half decades. The vast majority of chelated patients who suffered adverse effects experienced them during the 1960s, when chelation pioneers were still uncertain as to proper dose and administration of EDTA and much larger doses were administered in a much shorter time. Since then, chelation therapy has had a remarkable safety record.

It is possible, at worst, that over the decades twenty-five patients might have died, out of the more than one million treated, as a result of chelation improperly administered. That is a mortality rate of approximately one-thousandth of 1 percent. Compare that to the death rate from bypass surgery—up to 5 percent. Using the worst possible figures from chelation critics, compared to the best figures for bypass, chelation is a thousand times safer.

Which brings us at last to the real danger of chelation— the one not even its advocates are publicly acknowledging. The real danger stems from the potential for untrained, unqualified, unethical doctors jumping on the chelation bandwagon. Why now? Because every day chelation therapy is becoming better known and more sought after by growing numbers of people.

Recovered patients are outspoken salesmen. Even the tight-lipped are good advertisements—walking billboards. The more enthusiastic people become, and they have good reason for their enthusiasm, the more likelihood greedy opportunists will find ways to exploit the unwary. It is happening already.

The "quick-buck" boys are ready, waiting in the wings to cash in on the efforts of those who worked so hard and struggled for so many years to legitimize chelation and bring it to public attention and professional acceptance.

Chelation "mills," for the most part laymen-owned, have appeared in many of the more populated areas of the country and across the border in Mexico. They adhere to the supermarket concept of medical treatment, which holds the promise not of quality care for the patients, but of financial gain for the operators.

Entrepreneur types, with aspirations of establishing nationwide networks, have offered franchises to businessmen who can profit from chelation by hiring physicians to staff their premises, in some cases on a part-time, limited basis. Costs are kept to a minimum by employing doctors and staff unskilled and inexperienced in chelation. They hire public relations firms to design high-power advertising campaigns to promote their clinics via overstated ads that are excessively critical of conventional medicine. They urge patients to elect chelation for anything and everything that might ail them.

In one such syndicate operation, the doctor hired to oversee and supervise patient care at three chelation clinics sixty miles apart was allegedly untrained in chelation other than a brief period spent observing treatment procedures. Investigation of his medical background revealed a questionable history. He had previously been involved with setting up large, profit-making "emergency" centers in a way that had so outraged the community that he had closed them overnight and allegedly skipped town, taking the investors' money with him and leaving huge debts.

In the first two weeks of its operation, this sham clinic was reported to have one cardiac arrest that we know of, but the patient was resuscitated. As of this writing, there have been no deaths, but legitimate chelationists are holding their breath. In place of the careful monitoring by a well-trained staff, so vital to patient safety, patients are sometimes rushed through, with little or no individual attention, hooked up to overly rapid IVs, and turned loose. They are often given little or no nutritional supplementation, no dietary instructions, no instructions as to smoking, exercise, and the improved lifestyle changes that are an integral part of a complete chelation program. In one clinic I know of, patients routinely smoke cigarettes while receiving chelation therapy.

Obviously, this situation must be addressed and controlled or chelation therapy could be destroyed before universal acceptance has occurred. Chelation critics are quick to latch on to any excuse to feed disadvantageous tidbits to the sensation-hungry media.

Imagine their glee if Mike Wallace, Morley Safer, and the rest of the *60 Minutes* crew could zero in on a chelation center where fifty or one hundred patients per day were "processed" without adequate medical supervision. Imagine the public outcry and repercussions triggered by an exposé documenting a chelation center where patients were chelated too rapidly (to effect a more profitable turnover) or too frequently, regardless of whether they have full kidney function, one kidney, half a kidney, or are close to kidney collapse.

There are such places. One of my patients, who had been previously chelated in another state, complained to me after her first treatment at my clinic, "Why did it take so long, doctor? It went much faster where I was chelated before."

Despite well-established evidence that chelation infusions should be timed to last three hours or more, I have heard reports of sixty-minute treatments. Chelation "supermarkets" are setting themselves up for the full *60 Minutes* treatment (pun intended).

There are two real dangers at hand. The first is to the public at large—the danger of being treated by an incompetent or greedy chelator. The second is the possible end of chelation. There is no doubt chelation is in danger of being condemned because of the excesses of its high-profile exploiters and their blatant advertisements. The same popularity that might soon make it possible for victims of cardiovascular disease to have easy access to a safe and effective nonsurgical procedure could doom that very therapy before it has the chance to save limbs and lives.

The professional association of physicians interested in promoting the competent and ethical use of chelation is attempting to weed out unqualified physicians. It has established the American College for Advancement in Medicine (ACAM), which maintains standards of practice and recommends certification of physicians only after they have taken training courses, have passed written and oral examinations, and have had extensive, documented experience.

But this relatively small professional group cannot watch every aspect of the mushrooming chelation movement. For example, it cannot halt the aggressive marketing of so-called "oral chelators," being advertised to the public as capable of producing benefits similar to intravenous EDTA. Such products are mostly just overpriced vitamin-mineral supplements. Some do contain EDTA, but very little is absorbed by mouth, less than 5 percent. Most remains in the digestive tract, where it binds tightly to essential nutritional trace elements, blocking absorption, and potentially causing serious deficiencies. Recently I have even seen EDTA rectal suppositories advertised although research studies have shown that very little EDTA gets in by this route and it also has the potential to cause rectal hemorrhage and cancer. Despite the glowing testimonials contained in ads for these various substances, no oral or rectal chelator comes close to being a suitable substitute, and they are potentially dangerous.

ACAM has neither the resources nor the clout to establish a fail-safe chelation environment. An informed patient is his or her own best protector. To avoid hazards, know what to look for and what questions to ask. Be wary of any physician that does not follow the recommended protocol, using methods described in this book, which is the official accepted standard of practice in use by the American College for Advancement in Medicine.

Up to date information is periodically posted on my website: www.drcranton.com.

12

THE CHELATION EXPERIENCE

First timers have a pleasant surprise in store when they arrive for their initial chelation treatment. Even hospital phobics feel at home, reassured by the relaxed, informal, and friendly ambience characteristic of most treatment centers.

Unlike most other forms of treatment, being chelated is usually a group experience. Chelations are given in a comfortable, family-type room, furnished with cushioned easy chairs and recliners. You might think you had wandered into the lounge of a private club, were it not for each of the dozen or more people who are napping, chatting, reading, or watching TV connected to a container of fluid dripping slowly into an arm or hand.

Newcomers respond well to the chatty, light-hearted atmosphere. Because patients spend long hours together, they are soon on a first-name basis, swapping medical experiences, diet tips, and snapshots of grandchildren. They talk politics, sports, exchange family histories, and, most important, provide encouragement and support for new patients. Some patients prefer privacy, or a quiet atmosphere to read, and are provided a space apart.

You hear it so often.

"Don't worry. You'll do fine. Let me tell you the shape I was in when I began."

What's it like to be chelated?

"It's like attending a family reunion, where you like all your relatives," a patient once remarked.

Others have said, "It's like my weekly visit to the beauty salon," or "like having group therapy," or, more to the point, "like joining a society where everyone shares a very special, common interest."

What's true of the Mount Rainier Clinic, in Yelm, Washington (near Seattle), and the Mount Rogers Clinic in Trout Dale, Virginia—two clinics where I personally practice and consult—is also true of most other chelation facilities I have heard of or visited.

Ironically, chelation critics have tried to turn a plus—the pleasant surroundings and psychological support—into a minus. Many contend that patient recoveries can be attributed to the power of suggestion, as though chelationists were involved in some form of group hypnosis, bewitching patients into imagining themselves well.

Contradicting this oft-leveled criticism is a report from an outstanding expert in veterinary medicine who happens to use EDTA chelation therapy on racehorses. Dr. Lloyd S. McKibbon has observed remarkable changes in the racing performance of treated animals, verified by physiological monitoring tests. If chelation is really only a placebo, as many opponents would have people believe, how amazing that even racehorses can be fooled!

Indeed, were chelation specialists able to mesmerize people with blackened, gangrenous toes back to health with "think pink" suggestions, such feats, in themselves remarkable, would be worthy of acclaim.

I have no doubt that there are psychological benefits associated with chelation therapy. We do everything we can to encourage healing with lots of individual attention, group support, and hope.

Why such emphasis on the social environment?

Many chelation patients are very sick indeed, perhaps with a gangrenous extremity or at great risk of heart attack or stroke. Not infrequently, they've been advised to face the inevitable by physicians who've told them nothing more could be done.

Our immediate goal is to reduce their anxiety. What better way than to expose them to firsthand testimonials. That's not the only reason we encourage conviviality. It makes life pleasant for everyone, staff and patients alike.

As a patient once told me, "I hate being around sick people. My best surprise was to find that no one acted sick. Even those patients I later discovered had really serious conditions were joking and smiling. In all my weeks of treatment, I don't remember anyone who looked really grim or sad. I certainly never saw anybody crying."

And a more practical lady remarked, "The worst part of being chelated is it's so boring. I'd go nuts if there weren't people to talk to."

Not that being chelated is all fun and games. It is, after all, a biochemically complex treatment. While it appears simple to the uninitiated, behind the scenes a highly sophisticated and carefully planned protocol regulates the course of individual treatments.

To give you a clearer understanding of the chelation experience, from both sides of the physician's desk, let's follow a mythical new patient step by step. We'll call him Mr. B. He's fifty-seven, has had a well-documented heart attack (myocardial infarction), and he takes a full bottle of one hundred nitroglycerin tablets every month, which only partially relieves his angina. He can't walk more than one hundred yards with a small bag of groceries without stopping to catch his breath.

When he first called the office to inquire about chelation, Mr. B. was mailed an information packet.

On his first visit, he arrived prepared with a long list of questions typical of those most often asked about the procedure.

Where do we start?

We begin with a thorough pretreatment examination and laboratory testing. We start by obtaining a detailed medical history. We want to know all about you, not just your age, marital status, occupation, and previous medical problems, but also about your family's medical history. You will be asked questions about your diet, your lifestyle, your good and bad habits, whether you drink or smoke and how much, what doctor-prescribed or over-the-counter medications you take, what nutritional supplements you use, how much you sleep, play, and exercise.

We will ask about symptoms—pains, headaches, dizzy spells, unusual fatigue, swelling of the hands or feet, palpitations, nervous spells, heart flutter, everything you can think of to fully answer the question "How do you feel?" And we will arrange to get copies of pertinent past medical records from hospitals, doctors, or other available sources.

That's just for starters.

Next we do a complete hands-on, head-to-toe physical examination. We look in ears, eyes, nose, throat; we check for the regularity of pulses in neck, groin, and feet; then, using a stethoscope, we listen to those same pulses for bruits—with a stethoscope. Bruits are sounds made by turbulent blood flow past arterial plaques.

We check fingernails and earlobes for color, listen to breathing. We examine the chest, listen to the heart, get blood pressure readings, and do noninvasive vascular testing, such as the Doppler ultrasound, which measures blood flow adequacy in various parts of the body.

You'll be given an electrocardiogram, then a series of blood tests to ascertain the status of your biochemical health and determine whether you have diabetes, elevated blood fats, abnormal metabolism, liver disease, kidney disease, anemia, infection, a compromised immune system, or other problems. You will be asked to collect a timed urine specimen, which will be tested for heavy metal toxicity.

Whew! How long does all that take?

Usually half a day—anywhere from four to five hours. It may then take another day or more to receive the complete results of all the biochemical tests. Some doctors have testing instruments in the office for immediate test results of kidney function. In that case, it may be possible to receive your first chelation that same day.

What's next?

Assuming the examination reveals that your health status could be improved with a course of chelation therapy and you have no contraindicating conditions, a treatment schedule will be recommended, to be adjusted periodically to your progress and individual tolerance.

Many of my patients come for chelation while they are still totally healthy. They want to prevent problems in later life. Cardiovascular disease and cancer cause 75 percent of death and disability in this country, so it makes excellent sense to get preventive chelation, rather than take a chance on a stroke or heart attack later on. I always prefer preventive medicine to crisis intervention. And, although there is good evidence that chelation acts to prevent cancer, there is no evidence that it is effective once a malignancy becomes established.

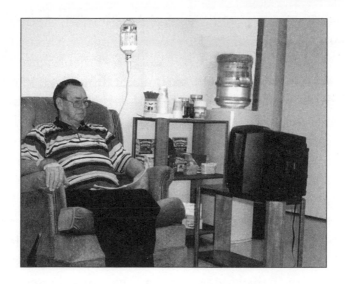

Chelation therapy is a slow process, but therein rests its safety. Most patients read, chat, doze, or watch TV. A light snack and fruit juice are customarily taken during treatment. This gentleman enjoys a video movie during his treatment.

I want to know exactly what to expect. What is the procedure? How is it done? What goes in the bottle? How will I feel?

After you are comfortably seated, a trained technician will begin your infusion. The prepared bottle of EDTA solution will be individualized for your weight and rate of excretion. A computer program is used to compute the correct dose of EDTA, so that all patients experience the same blood level throughout the infusion. The IV bottle will be hung near ceiling height on an adjustable stand or hook, and will be attached to a vein in the back of your hand or forearm via tubing ending in a tiny needle, which will be inserted as painlessly as possible. Most find the momentary pinprick a minimal discomfort. To keep the infusion set firmly in place, we sometimes attach the arm to a padded arm board, which then rests on a pillow. This provides greater comfort for the patient and allows him or her to move about more freely, without fear of dislocating the needle.

The technician will next adjust the drip rate as individually indicated. In most cases, it is timed to approximately one drop per second for a three-hour treatment. To make certain the infusion is being well tolerated, you will be carefully observed during the treatment session.

Prior to the treatments, we obtain periodic urine and blood specimens to ensure that kidney function is adequate to excrete the EDTA infusion safely. If a potential for overload is detected, or if there is a hint of kidney insufficiency, treatment is slowed or delayed, until the kidneys recover their normal reserve.

This continual checking procedure is especially important in elderly patients, whose kidneys can decline with age. Properly monitored, kidney safety can be assured. In most cases, kidney function actually improves after a course of therapy.

As for the chelating formula itself, the infusion normally consists of 500 milliliters (about one pint) of intravenous fluid containing EDTA, to which is added (according to individual needs) varying doses of vitamins and magnesium. We generally add B-complex vitamins, B-12, folate, and vitamin C, which is also a weak chelating agent that enhances EDTA's benefit. Magnesium, also beneficial for cardiovascular disease, is a common additive. We also include a tiny bit of heparin, which is an anticoagulant—not enough to thin the blood or prolong clogging time, but just enough to prevent a blood clot at the injection site. To prevent pain during the infusion, we add a local anesthetic—lidocaine—the same substance your dentist uses. We don't want it to hurt, and it shouldn't. Research in Europe shows that lidocaine alone has beneficial effect against atherosclerotic arterial disease.

Now, as to how you will feel. Everybody is different. Restless people and those who are easily bored should bring along something to busy themselves with: an entertaining book, study material, or a handiwork project, especially if they will not be content chatting, napping, or watching TV for several hours. A videocassette movie can also help the time pass more quickly.

Patients must remain comparatively immobile while being treated. They can move about gingerly if they must, to go to the bathroom or take a telephone call, but for the most part it is best to stay in place to prevent disruption of the infusion.

The vast majority of patients consider each treatment a pleasant respite, especially if, as is quite common, they begin to feel beneficial effects after a few sessions. Once the treatments start to take hold, many people can breathe, walk, and work without discomfort, after years of progressive deterioration, and they usually look forward to being chelated.

What if I have to interrupt the course of treatment? Or move away? Or just decide I want to quit?

You can stop treatment at any time and still retain the benefits from all sessions up to that point. Benefits accrue in direct proportion to the number of treatments received. Unless you wait one year or longer between treatments, benefits start to add up again with each new treatment.

There is no rebound effect with chelation or worsening of symptoms merely because treatments are stopped. Once you start, you do not have to be chelated periodically unless you want to. The schedule itself is not very important, only the number of treatments. The more frequently they are received, the sooner the benefit. It does no harm to stop for a month or so for vacation or business travel and then pick up where you left off on your return.

If kidney function remains normal, some patients receive as many as five treatments per week. Others come once per month. As long as the condition being treated is stable or improving, less frequent treatments still help but require more time to do so.

You can start treatment with one doctor and continue with another in some other part of the country, usually with no difficulty. The treatment protocol might vary a bit, but it should not be significantly different. The main ingredient is the EDTA. If EDTA gets into a vein, benefit should occur. The other ingredients add somewhat to benefit, but vitamins are also effective by mouth.

And what about side effects?

Although once in a while someone encounters a minor medical problem during chelation, chances are you won't. Among the possible side effects are the following:

• The EDTA infusion might cause minor irritation—a burning or stinging, or slight swelling, or a black and blue mark at the puncture site. An occasional patient develops a skin rash because of an allergic reaction. When necessary, we sometimes alter the ingredients to eliminate side effects.

• The therapy may upset your stomach, even make you feel slightly nauseated, a problem readily treated with medicine or vitamins. This adverse effect is experienced by less than 1 percent of all patients.

• An occasional patient develops a headache. If so, simple nonprescription pain medication relieves it.

• Once in a great while, a patient experiences sudden feelings of faintness, weakness, light-headedness, extreme fatigue, or dizziness if he stands up suddenly during or after treatment. This may occur because blood pressure can become a bit lower during treatment. In fact, many patients with high blood pressure have their antihypertension medication dosages decreased after chelation therapy. If that should occur, the immediate solution is to rest for awhile after treatment, with feet elevated, head lowered, until blood pressure has normalized.

• If you are hypoglycemic, you may experience a weak feeling or feel a bit "woozy" during chelation because blood sugar levels may drop. A glass of fruit juice or, better yet, a snack during treatment, will minimize the hypoglycemic reaction. This is especially important if you have diabetes and use insulin or oral medication to control blood sugar.

• One out of an estimated five hundred patients develops an unexplained fever within the first twenty-four hours after chelation. It usually disappears spontaneously without treatment but should be brought to the physician's attention.

• Approximately one in twenty patients develops temporary leg cramps, usually at night. EDTA does affect calcium-magnesium balances, and additional magnesium or calcium tends to ameliorate this problem.

• One percent of patients develop a loose stool or diarrhea, either immediately after treatment or the next day. This is remedied by antidiarrhea medication. More common is the need to empty the bladder more frequently. Some people report they have to get up once or twice that night. EDTA can act as a mild diuretic for a few hours after infusion. Chelation therapy improves tissue integrity, decreasing edema, returning accumulated fluids to the circulatory system to be excreted. That's why many patients experience an unexpected reward—a three- to five-pound weight loss of unwanted fluid retention.

• Diabetic patients must have their blood sugar carefully monitored to prevent inadvertent hyperinsulinism. Patients who use
insulin will usually notice that blood sugar is easier to control
and that insulin doses can be gradually reduced, and occasionally eliminated, during and after a course of chelation therapy.

How many treatments? Over how long a period of time?

We won't have the answer to those questions until after the
results of your medical examination, and even then we can
only estimate from experience. It's a highly individualized
matter. Patients being chelated preventively, prior to the onset
of disease symptoms, normally require fewer treatments:
twenty to begin with in the first year, and then five or six
yearly thereafter to maintain benefits. In cases where there is
an established history of atherosclerosis, angina attacks, or
other signs of arterial disease, we usually prescribe a series of
thirty treatments over a three- to six-month period, after
which we recommend one follow-up treatment every month
or as needed to maintain improvement.

Can you guarantee results?

Of course not, even though chelation has an enviable
record. While better than 75 percent of all patients treated
experience a reduction of symptoms related to arteriosclerotic-
type ailments, it is no panacea for all health problems that
accompany the aging process. It does not "cure" old age.
Sooner or later, human mortality catches up with us, chelated
or not. We can usually turn it back some and then slow it in
the future, but we can't stop the aging process entirely.

Chelation delays the inevitable. It helps most people live
longer and healthier, but only to the extent that they are willing
to modify their lifestyles. Once your genetic limit of age is
reached, somewhere between 85 and 120, depending on inherited traits, we have nothing at present to extend that cutoff point.

For full benefit, patients must quit smoking. Those who continue to use tobacco in any form jeopardize their chance of benefit. I often tell my patients that to continue to use tobacco while undergoing chelation is analogous to pouring kerosene on a fire while calling for the fire department to put it out. Smoking does not prevent benefit, but the improvements will not be as great and will not last as long. Some patients find it impossible to stop all tobacco. I do not refuse them chelation, but advise them that they will not get as much for their time and money.

Failure to be physically active may also decrease potential benefits. It is equally important for patients to adhere as closely as possible to recommended nutritional guidelines. It is not necessary to be fanatic about diet, but just use common sense, as explained in detail in chapter 14.

How will I know chelation is working?

Seriously ill patients, those with advanced forms of atherosclerotic-related ailments, rarely have reason to ask that question. Day by day, they enjoy an improved ability to function in a more normal way. A typical illustration is Mr. D., who, prior to chelation, could barely make his way from his living room couch to the dining room table. Six months after treatments began, he retiled his roof single-handedly in below-freezing temperature.

Mr. L., an electrician, was only in his mid-forties when the circulatory problems in his lower limbs started to make his life unbearable. Was it wheelchair time? After two angioplasties and more surgery recommended, he could hardly walk a block. Walking and standing to do his work was torturous. The climax for Mr. L. came when he got a call to help fix a shutdown in an electric power plant. The parking lot, located across a highway, was connected to the plant by a pedestrian bridge. Mr. L. had to get his coworkers to carry him up the stairs to the bridge. He took sick leave the next day.

A few months later he underwent his first chelation. By the fifth treatment he noticed tingling in his feet, where before they had been numb. By the twentieth treatment he was ready to return to work. Two years have passed and Mr. L. is working five and six days a week with no further leg or foot pain.

Patients who are chelated preventively will see more subtle improvements than those that thrilled Mr. D. and Mr. L. Skin takes on a more youthful glow, partly because of improved circulation and partly because of gradual reduction in the wrinkling, as cross-linkages caused by free radical damage are reduced. Hair condition—color and texture—often improves. That's because increased levels of antioxidants and free radical reduction by EDTA can sometimes help to stop hair loss and occasionally cause gray or white hair to return closer to its original color.

One more proof that chelation is doing its job: Those telltale brown spots, especially on the back of the hands (aptly called age spots), begin to fade away. Those same free radical reactions that cause deterioration of skin, tissues, arteries, and organs throughout the body also give rise to yellow-brown frecklelike marks, also called "liver spots" (most noticeable on backs of the hands). These are visible evidence that cellular wastes are building up throughout your system. Over a period of several months, chelation slowly reduces those cellular wastes, and the ugly age spots gradually fade, a far better solution to the problem than widely advertised commercial "fade-liver-spots creams" designed to "bleach" the skin clear.

I've run out of questions.

Let me answer the one important question you failed to ask: How can you be assured of a chelating physician's competence?

As with many other medical procedures, you should investigate the knowledge, experience, and credentials of the physician you select.

To protect yourself, ask if the doctor you are considering has completed training courses taught by doctors who are expert in chelation therapy. Don't be bashful. Ask, and above all, don't sign a treatment consent form unless you're convinced you're in safe hands.

To protect yourself against being victimized by a "chelation mill," watch for these telltale omissions:

• Failure to obtain a complete medical history, including pertinent hospital records, recent x-ray reports, EKGs, arteriograms, or other lab tests.

• Failure to do a thorough hands-on physical examination, if not done and recorded recently elsewhere. This should include a one-on-one inquiry into such lifestyle habits as eating, smoking, exercise, and drinking, a comprehensive series of blood tests, a urine collection, and some form of noninvasive vascular testing.

Once chelation treatments have begun, there are other warning signs to watch out for:

• Failure to periodically monitor kidney function with urine and blood tests during treatment.

• Failure to supply information about test results or answer questions about procedures, symptoms, or progress.

• Failure to give personalized, individual attention.

No treatment is any better than the physician administering it. Unfortunately, there are quack chelationists around, just as there are opportunists in every other medical specialty.

One chelation doctor I heard of claimed to be using a patented secret formula not available to other doctors, for which he charged two to three times the going EDTA chelation fee. Subsequent analysis revealed he was using the same EDTA as everyone else.

Your best protection is to be an informed patient. When your doctor balks at answering questions, it might be time to seek another physician.

The important ingredient in chelation therapy is EDTA. If the doctor gets an adequate dose of EDTA into your veins, you can expect benefit. Other types of therapy may be recommended, but a full course of EDTA is the most important part of therapy.

This lady enjoys a good book during her chelation treatment. The infusion bottle hangs above and is connected by tubing to a tiny needle in her arm. She can stand and stretch her legs from time to time, or carry the infusion bottle with her to the restroom.

13

YOU HAVE OTHER ALTERNATIVES

What are the alternatives to bypass surgery? The day it could be crucial for you to know, you might not have time to find out.

Consider what happened to forty-eight-year-old Richard J., an advertising agency executive, when he, like a million and a half other American men and women each year, had a heart attack without warning.

Richard had just finished lunch with a client and was heading back to the office when terrible chest pains struck. Overweight, overstressed, underactive, a two-martini lunch man, and a smoker, Richard suspected the worst as the ambulance rushed him to the hospital.

When his wife, Alice, arrived, doctors were ready with a diagnosis—advanced atherosclerosis with three-vessel coronary disease—and a remedy, immediate bypass surgery.

The cardiologist tried to be reassuring.

"Your husband's a lucky man," he said. "He's had a heart attack and is under sedation, but he's doing okay. We've located his trouble. These films show the arteries that are blocked."

"What now?"

"We've scheduled surgery for first thing tomorrow, but he's in no condition to okay the operating permit. Sign here."

Mrs. J. vaguely recalled reading something about bypass surgery but could not remember whether it was good or bad. She thought of Dick's friends, his golf buddy who'd died on the operating table and the art director who'd had his bypass five years ago and who looked and felt better than ever before.

"Right away?"

"There's no time to fool around," cautioned the surgeon. "Your husband's a walking time bomb. He could live forever or die any minute. Those clogged arteries are keeping oxygen from reaching his heart."

What would you do? Would you stand your ground and insist on more time? Would you swallow your doubts and hope for the best?

Given all the facts and an opportunity to weigh the comparative risks versus benefits of surgery and other invasive treatment programs (as detailed in chapter 16), I have no doubt most people would bypass the bypass and take their time making a fully informed decision.

But they are not often given the chance. In one instance, the surviving family of a man who had died during a bypass operation struck back. They instituted a class-action lawsuit against the university hospital, alleging the heart surgery booklet they had been given concealed the hospital's excessively high mortality rates, thus negating the validity of the operating permit because it was not signed with "informed" consent. In a similar case, the family sued because they were not told about chelation until after their dad died during bypass surgery.

Is a physician bound to inform you of all alternatives, those he disapproves of as well as those he endorses? Ethically? Yes. Legally? Not always, although laws in some states do require such full disclosure.

Some doctors think modern medicine is too complicated to discuss with the average person.

Some doctors are so steeped in the jargon of their medical mystique that they are unable to describe things in understandable, simple English.

Some doctors appropriate the decision-making responsibility, assuming patients and their families, traumatized by a medical emergency, are too confused to think for themselves.

Some doctors are so prejudiced in favor of their own specialty that they disregard the potential for greater benefit from a competitive therapy.

Some doctors have simply failed to keep up with medical advances and new research.

And, let's face it, some doctors have too great an investment in their own specialized skills to be interested in therapies that require retraining.

Nevertheless, in the last fifteen years, medicine has taken some giant steps toward the conquest of heart disease. There's no better way to document progress than to realize that some hundreds of thousands of Americans are living today who would have died a decade ago were it not for scientific breakthroughs on many fronts—from prevention to cure. Today doctors know more about helping people get well, and people know more about keeping themselves healthy.

To maintain control over your life, even when faced with a serious health reversal, do not give up your decision making prerogatives. Ask questions, be well read, keep an open mind, and inform yourself.

What are your alternatives if you reject bypass surgery? Are you in that small group of patients for whom the risk of surgery is less than no surgery (as defined in chapter 16). You've already learned a great deal about chelation therapy; now let's look at other possibilities as well.

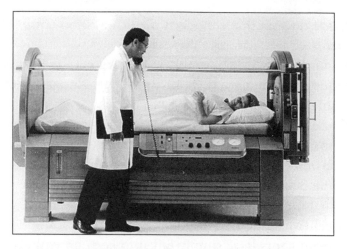

This patient is receiving hyperbaric oxygen therapy (HBO) in a Sechrist monoplace chamber. At two atmospheres of pressure in 100 percent oxygen, body tissues experience a tenfold increase in oxygen. Full view and two-way communications between the patient and those outside the chamber are maintained throughout the customary one-hour treatment.

Hyperbaric Oxygen Therapy

Hyper means "increased" and *baric* means "pressure." Hyperbaric oxygen therapy (HBO) involves intermittent treatment of the entire body with 100 percent oxygen at twice normal atmospheric pressure. Our surrounding atmosphere contains approximately one-fifth oxygen. One hundred percent oxygen will therefore increase that amount by five times. One hundred percent oxygen at twice normal atmospheric pressure increases oxygen exposure of the body by tenfold.

While some of HBO's mechanisms of action are yet to be discovered, it is known that HBO (1) greatly increases oxygen concentration in all body tissues, irrespective of blood flow; (2) stimulates the growth of tiny new blood vessels to areas with reduced circulation, providing collateral circulation around arterial blockage; (3) causes a rebound arterial dilation, with increase in blood vessel diameter greater than when therapy began, improving blood flow to compromised organs; (4) stimulates an adaptive increase in superoxide dismutase (SOD), the body's internally produced free radical scavenger; and (5) greatly aids the treatment of infection by enhancing white blood cell action, potentiating germ-killing antibiotics.

While not new, HBO has only lately begun to gain recognition for the treatment of chronic degenerative health problems related to atherosclerosis, senility, stroke, brain damage, peripheral vascular disease, and other circulatory disorders. Although oxygen delivery to vital organs is reduced by atherosclerosis, symptoms of occlusive artery disease, such as incipient gangrene, can be rapidly reversed with HBO.

Studies show that two hours daily of simply breathing pure oxygen, even without a hyperbaric chamber, may speed artery healing. Chicago's Dr. Vesselinovitch demonstrated this beneficial effect of breathing oxygen with atherosclerotic rabbits who were given an "oxygen session" each day. Like many other things, however, because a little is good does not mean that more is better. Oxygen therapy beyond a safe limit can be toxic.

One of the world's most experienced authorities on hyperbaric medicine is Dr. Edgar End, clinical professor of environmental medicine at the Medical College of Wisconsin, who voiced his opinion on HBO's value for the treatment of stroke this way: "I've seen partially paralyzed people half carried into the (HBO) chamber, and they walk out after the first treatment. If we got to these people quickly, we could prevent a great deal of damage." Even when HBO is administered years after a stroke, patients have experienced improvement of function.

HBO is usually administered in a transparent, heavy-gauge acrylic cylindrical chamber about eight feet long and three feet in diameter. The patient, made comfortable on a foam mattress, is rolled into the chamber, which is then sealed shut. While in the chamber, which is equipped with microphones and speakers, the patient can watch TV, listen to the radio, read, nap, and talk with the chamber operator or relatives is outside. During therapy, usually lasting one hour, the patient is surrounded by and inhales pure oxygen, while pressure within the chamber is increased to twice the outside pressure, equivalent to what a diver would experience at approximately thirty feet below the surface of the water. At the end of the treatment, the patient is gradually decompressed to normal pressure.

It is only in the last two decades that HBO has come to be used in conjunction with chelation therapy. Results have been exciting. Patients with cerebral vascular disease recover from complications of strokes more readily when these two treatments are combined. Senile patients, including those with Alzheimer's type, respond better to both therapies than either one alone. The same holds true for gangrenous legs and feet caused by blocked circulation. HBO rapidly relieves pain, helps eliminate infection, and keeps the threatened tissues alive while chelation therapy gradually improves circulation. Recently, HBO has been successfully used to help patients with cerebral palsy and following brain trauma.

The main drawback to HBO is the difficulty of securing treatment due to the shortage of HBO facilities with chambers. There are estimated to be only about one thousand in the world, and many are in Russia. Of those in the United States, many are huge, operating-room size, multiplace chambers, which require an engineer and a crew of technicians to operate, making them so inefficient and expensive that they are not very practical.

Drugs

Long-term therapy with prescription drugs can be very helpful when the basic health problem is angina, arrhythmia, hypertension, or other conditions related to impairment of coronary artery blood flow.

Drugs that might be recommended include the following:

Anticoagulants: Primarily useful in diseases of the veins, such as thrombophlebitis, where there is a danger of blood clots forming and breaking loose.

Anticoagulants are blood thinners (the best known is Coumadin or wafarin) and must be carefully used lest they precipitate hemorrhage. This is clearly a potential hazard since the most common of these drugs, warfarin, is found in hardware stores, where it's sold as rat poison. When rats eat enough, they hemorrhage and bleed to death internally.

Anticoagulants are rarely appropriate for patients with angina or atherosclerotic diseases unless the danger of a blood clot embolizing from the wall of the heart is greater than the danger of possible hemorrhage as a result of the therapy. If your physician recommends such drugs, you might want a second opinion. Some patients, but not all, with a rhythm disturbance called atrial fibrillation should definitely be on anticoagulants.

A less hazardous approach to decreasing the likelihood of blood clots, but with only slight benefit, involves aspirin or other drugs that decrease platelet "stickiness." There is some evidence that small daily doses of aspirin may reduce the risk of heart attacks and strokes. Vitamin supplements, including vitamin E, also help to prevent abnormal clotting.

Another approach to this problem is a change in diet (described in chapter 14). But heed this warning:

If you are currently taking anticoagulants, do not go on the anti-free radical diet or make any changes in your food habits or take aspirin or any other new medicine without consulting your doctor. If you do, what may have been a safe dose might become a fatal dose. Use of anticoagulants requires regular blood testing for prothrombin time, a measure of clotting ability. Any change in diet, medications, vitamins, or lifestyle can alter the correct dose.

Blood vessel dilators: That old standby, nitroglycerin, is still used and is still appropriate for relief of angina. It, like more recently developed short- or long-acting nitrates, reduces spasm in blood vessels and decreases the workload of the heart. Angina can be relieved when the heart does not have to pump so hard or if coronary artery spasm is reversed. While drugs in this category do nothing to treat the underlying disease, they can, in some cases, reduce the risk of heart attack and relieve symptoms.

New forms of this old drug include adhesive skin patches impregnated with nitroglycerin as well as a nitroglycerin skin ointment. Slow-release nitrates may be useful prior to exertion such as lovemaking or walking.

While the nitrates can be very effective, either as a stopgap measure or for long-term use, they have their dark side. Among the more commonly observed adverse reactions are headaches, transient episodes of dizziness and weakness, and occasional skin rashes. If their use is interrupted without a gradual tapering off, angina or even a heart attack might be precipitated.

Beta-blockers: Now marketed under dozens of trade names (Inderal, Blocadren, Tenormin, Lopressor, Atenolol), these drugs work on the body's adrenergic (adrenalin related) nervous system, which is involved in stimulating the activity of the heart and blood vessels in stressful situations. They interfere with the transmission of nerve impulses that trigger beta-receptors, allowing relaxation of arteries and reducing the heart's oxygen requirements.

There is now substantial evidence that beta-blockers can reduce the incidence—and severity—of heart attacks in some patients.

New and more recently synthesized beta-blockers are expanding their usefulness. Used either alone or in conjunction with other medicines, they can effectively control both angina and hypertension.

A substantial percentage of patients experience a variety of unpleasant side effects (not as often with the newer beta-blockers): sleeplessness, bad dreams, lowered sex drive, reduced energy. As one patient put it: "My get up and go, just got up and went."

Some beta-blockers can present problems for asthmatics and people with respiratory problems by constricting air passages in the lungs.

Calcium channel blockers: By interfering with the flow of calcium into heart and arterial muscle cells, calcium blockers reduce the intensity of the heart's contraction and may reduce rhythm disturbances. Less intense contraction allows the heart to rest slightly, reducing its need for oxygen. At the same time, they can reduce spasm in the coronary arteries, increasing the blood flow and oxygen supply to the heart.

Also called calcium antagonists, these drugs have been used successfully worldwide for more almost three decades. Doctors have found them amazingly helpful in reducing angina, treating heart rhythm disturbances, relieving hypertension, and reducing heart attacks.

A powerful group of drugs, they seem to be safer than many other medications, with fewer negative effects in most patients. Some calcium blockers are proving as effective as beta-blockers in the treatment of high blood pressure, with the advantage that they do not affect the lungs.

With the ability to reduce the damage that occurs when calcium leaks into cells where it does not belong, calcium blockers duplicate just one of the many therapeutic actions of chelation. The difference is that chelation results in long-term benefits after a course of infusions, while prescription drugs must be taken continuously.

Cholesterol-lowering or lipid-lowering substances: Medications in this category, much like low-cholesterol diets, put the cart before the horse inasmuch as they are aimed at attacking the symptom rather than the problem. They can occasionally cause nasty side effects, including cataracts, impairment of liver function, and an increased risk of gallstones and cancer. These substances may be indicated, however, for some patients with a hereditary tendency to very high blood cholesterol. Some experts now think that benefit from cholesterol-lowering drugs is related to their antioxidant and antiplatelet activity and that nutritional supplements are better in that regard.

Digitalis: Under certain conditions, digitalis may relieve heart arrhythmias. Under other conditions digitalis may produce heart arrhythmias. In certain patients, digitalis may relieve angina. There is an equal possibility that it can aggravate angina.

The role of digitalis in treating coronary heart disease is very speculative (it may relieve symptoms of heart failure); it should not be used routinely. If your doctor prescribes it, you may want a second opinion. The toxic dose of digitalis is so close to the effective therapeutic dose that potential benefits must outweigh the risk. Periodic measurement of blood levels can ensure a safe dose for each patient.

Clot-dissolving enzymes: Streptokinase and other newer clot busters are materials produced either by strep bacteria in a culture medium and then purified or by newer biotechnology. They are clot-dissolving chemicals that, when injected, can either stop a heart attack in progress or eliminate existing clots in veins or arteries. They also reduce platelet stickiness and probably help relax arterial spasm as well.

To effectively stop a heart attack, clot-dissolving enzymes must be used promptly. The patient must be in the right place—close to a hospital—at the right time, with a trained physician standing by. In many cases, catheterization is necessary to locate the site of the clot. They may also increase the risk of hemorrhage. Treatments with clot-dissolving enzymes can be extremely valuable in heart attack emergencies but have no impact on the underlying atherosclerotic process involved in clot formation.

Every drug, misused, can be dangerous. Protect yourself. Seek the care of a skilled physician. Adhere to these dos and don'ts of proper drug-taking:

• DON'T stop taking drugs just because you are feeling better. Consult your doctor even if you no longer experience the symptoms for which the drug was prescribed.

• DO follow label instructions as to frequency, timing, and size of dose. Don't take a drug at random intervals or on an empty stomach if "with meals" is specified (or vice versa). Don't "double up" if you miss a dose without first consulting you doctor.

• DON'T drink alcoholic beverages while taking drugs unless your physician specifically says it is safe to do so.

• DO throw leftover or outdated medicines away promptly. Some become less potent, others dangerously more potent.

• DON'T store similar-looking pills in the same location or in same-color or same-size containers. Look-alike drugs are easily confused, especially at night or when you are tired, in pain, or under stress.

• DO be sure your doctor and pharmacist are fully informed of other medications you are taking—prescription or over-the-counter drugs—at the time the doctor prescribes a new drug. Be careful not to mix medications without his knowledge and approval.

• DON'T store drugs where they may be damaged by excessive heat, cold, light, or dampness. Avoid cabinets in bathrooms that frequently get hot and steamy.

• DO call your doctor immediately when you experience an adverse drug reaction, develop unusual symptoms, or unexpected feelings.

Surgeries Other Than Bypass

Great strides are being made in the development of competitive treatments to bypass surgery, many of them significantly less costly, simpler, and safer. Among the more promising advances:

PTCA (percutaneous transluminal coronary angioplasty): In this procedure (alternately called balloon angioplasty), a thin tube or catheter, with a tiny balloon at its tip, is snaked into a coronary (or other) artery, close to the area of a well-defined blockage. The catheter's progress is followed by direct visualization on a TV-like screen, so that it can be guided to the blockage. After the balloon is inserted through the area of plaque blockage, it is inflated, stretching the arteries. Since these blood vessels are not very elastic, they do not fully reconstrict, thus leaving a larger residual opening than was there originally, freeing up blood flow.

While there is some risk of plaque breaking off or of the stretching process causing tears in blood vessel walls, neither has proved to offset some potential benefit.

Angioplasty, the hot new successor to bypass surgery, also has limited benefits. It may be highly effective for localized blockages in accessible areas if such arterial blockages are large enough to accommodate the catheter. Like surgery, the process can only unblock small segments of a diseased artery and may prove useless in patients where the disease has progressed too far or where blockages are unreachable.

In recent years it has become common to place a stent (a wire or mesh cagelike device) inside the artery after balloon angioplasty to prevent reclosure. It's too soon to know how successful this procedure will be in the long term.

Laser surgery: Theoretically promising, this technique is in its infancy. But if all the technical problems can be solved, it has tremendous potential.

Once the procedure is perfected, laser beams may be delivered via a tube containing superfine glass fibers through the interior of clogged blood vessels, vaporizing obstructions directly, one by one. Researchers developing the process say they will be able to accurately and safely direct and adjust the intensity of the laser light to individual plaque requirements.

Although fiber optic surgery is still in the dry run stage, fiber optic experts are confident that lasers will be the surgical tool of the future and expect to be vaporizing plaque clinically very soon. However, the problem of downstream embolization of plaque debris must also be solved. If so, laser surgery could eventually be a relatively minor procedure, administered on an outpatient basis. But it would still share the same drawbacks as angioplasty—useful only for readily accessible plaque.

Other roto-rooter-like procedures that use rotating blades to ream out plaque are also being researched.

Two other surgical procedures should be mentioned although they do not counteract atherosclerotic disease:

Pacemakers: There is a world of difference between the first crude devices implanted more than four decades ago and the pacemakers some half-million Americans depend on today to provide regular electrical impulses and prevent their hearts from beating irregularly (too fast, too slow, or not at all). The newest versions of these life-saving devices, inserted beneath the skin of the patient's chest, with wires reaching to the heart, need little attention. They work on tiny, powerful, lithium batteries that last for five or more years and can almost think for themselves. They contain a tiny computer that adjusts automatically to heart rhythm malfunction and changing body demands.

For the small percentage of heart disease sufferers who require a pacemaker, it is a remarkable invention but has little to offer the vast majority of persons with atherosclerosis.

Heart replacement: Heart transplants were all but abandoned in earlier years, when most recipients survived only briefly. The operation was usually a success, but the patient died within weeks or months nonetheless, either because of organ rejection or massive infection. Today, newer drugs, better tissue-matching techniques, and improved rejection monitoring procedures have allowed the survival rate to increase to more than 80 percent. This is now a viable option for some patients.

When one is caught up in the euphoria of Buck Rogers medical achievement, it's easy to overlook reality and the more dismal and sobering facts.

Donor organs are hard to come by. There are a limited number of young, healthy hearts not injured by a death-causing accident. (An interesting sidelight: Early in heart transplant research, surgeons discovered they were unable to keep donor hearts alive and well unless they were immersed in a solution containing EDTA.)

Transplant enthusiasts expect to overcome the organ shortage problem by encouraging more public participation in donor programs, by perfecting xenografts (transplantation of organs from such animals as pigs, cattle, and sheep), and by developing functional, practical versions of the artificial heart.

Even if all that happens, will heart replacement then be a feasible alternative for masses of people?

Let's talk money. Should heart replacement technology be perfected, the costs—$75,000 to $150,000 per procedure plus $8,000 to $20,000 per year for ongoing care—would put it in the break-the-bank category, whether privately or publicly funded.

Let's assume that financial drawbacks can be overcome. Then can the average American hope to live longer by virtue of perfected new-hearts-for-old technology? Not even then.

Remaining is the drawback faced by most heart replacement recipients. They are stuck with the same old diseased arteries. The new hearts, real or artificial, may be capable of beating on indefinitely, but what use are they when patients have not been cured of their underlying disease? Coronary arteries in the new heart are once again afflicted with atherosclerosis.

The $200 million or more thus far devoted to the development of transplant technology has been badly misspent, inasmuch as it is a piecemeal attack on the problem. Atherosclerosis affects the entire body, not just one organ, or one segment of artery. Heart replacements, like coronary bypass surgery, neither correct the precipitating ailment nor reverse its progress.

Patients with a disease called cardiomyopathy, an isolated paralysis of heart muscle without atherosclerosis, may be the best candidates for heart transplant surgery.

Lifestyle Changes

Improved diet, regular exercise, nutritional supplements, stress management, and avoidance of tobacco and excessive alcohol, either alone or in combination with other therapies, have proven benefits. Most thoughtful researchers recognize that advances in treatment for heart disease account for only a small part of the significant decline in heart-related deaths in recent years. Changes in health habits may be even more important. Record numbers of Americans have become "health nuts." They're jogging, eating better, and buying nutritional supplements. As scientifically described elsewhere in the book, many vitamins, minerals, and trace elements are essential to the safe control of free radical reactions in the body. Vitamins C and E, beta-carotene, and other anti-oxidants are free radical scavengers. Anything that will slow or control free radical damage is certain to slow the progress of degenerative disease. (More detailed information regarding diet and nutritional supplementation is contained in chapter 14.)

Most cardiologists, whatever treatment they prescribe, prod their patients into adopting better health habits. My experience tells me that those patients who do adopt better habits undoubtedly do better, and thus they support the success rate of whatever therapy they're on.

All qualified chelation doctors consider lifestyle changes an integral part of the entire chelation program—so much so, that in my practice, acknowledgment is written into the patients' informed consent literature.

In this category, there are lots of "right" things to do, and none of them will hurt you. While the evidence for benefit is not fully established, that is small reason to wait for all experts to agree on what may save your arteries—and your life.

To emphasize how much importance we place on lifestyle changes for their potential to reverse as well as prevent disease, we are devoting two entire chapters to practical ways of modifying health habits.

One such approach deserves special recognition. Advocated by both the Pritikin Program and the Dr. Dean Ornish programs, this approach incorporates diet and exercise and involves drastic reduction in dietary fat.

Hundreds of thousands of cardiac patients now believe in, and follow, a low-fat diet. Early claims of success are now supported by numerous published studies and testimonials from men and women, many of whom canceled their scheduled bypass surgeries to give the diet and exercise regimen a try.

It is not easy to stick to such a program. The diet, a very Spartan, high-complex-carbohydrate, low-fat diet that allows patients only 10 to 15 percent of daily calories as fat (the average American diet contains 40 percent of calories as fat), requires extraordinary discipline to abstain from your favorite "goodies," to give up smoking, and to exercise extensively.

Such a program forbids eating all of the following: fats and oils, sugars, fatty meats, fried foods, shrimp, egg yolks, all but nonfat dairy products, jams and jellies, nuts, dried fruits or sweetened fruit juices, commercial products made of white flour or white rice, soda pop, coffee, tea, and alcohol.

As a reward, adherents to the program are offered the hope they will recover their health, and be pain- and symptom-free.

No matter how beneficial it has proven for some, an extremely low-fat diet is not for everyone. It is so exceptionally restrictive, with such rigid dietary restraints, that only a small percentage of people can sustain the cultlike devotion required for long-term compliance.

I've come to believe that the quality of food selected, combined with nutritional supplementation, is more important than the highly individualized ratios of fat to protein to carbohydrate (discussed further in chapter 14).

Nevertheless, I believe we owe a debt of gratitude to Nathan Pritikin and Dr. Dean Ornish for popularizing a general approach we might all be well advised to follow, though in a modified form.

Fat and oil in the diet are harmful only when oxidized. I believe that by avoiding rancid and hydrogenated fats and oils, combined with a scientifically balanced spectrum of nutritional supplements, we can largely prevent adverse effects of prudent amounts of dietary fat. I have no objection to consumption of 30 percent or even more of calories as unadulterated fats and oils, if the proper nutritional supplements are taken on a daily basis. In my opinion, if Nathan Pritikin and Dr. Dean Ornish had incorporated high potency and broad-spectrum, vitamin-mineral supplements into their respective programs, results would have been as good or better with much less Spartan restrictions.

And there is one final alternative:

No Treatment

Rest and relaxation (R&R) may be all that is required for some people to go on living much the same as they did prior to their heart attack or angina pains. We have no statistics, of course, on people who opt for the hands-off approach to their disease, but I suspect that some few may do as well as those who pursue the most aggressive avenues.

While I am clearly biased in favor of the therapies I know best and have seen work so well for so many patients, I have no quarrel with people who choose to use other therapeutic means. Furthermore, I have no objection to surgery where clearly indicated and appropriate, and I continue to refer such patients for bypass or angioplasty. But in my view, bypass surgery should be reserved in most cases until all nonsurgical therapies have been tried and found wanting. It should be the alternative of last resort.

14

Anti-Free Radical Prolongevity Diet

The diet you'll be eating to avoid cardiovascular ill health will—for almost every one of you—be the same diet that you would design to ensure yourself a long and healthy life. That's because most good diets have general principles in common. And they have certain long-term results in common, too. Most of you will find that eating right keeps you looking trim and vigorous, with healthy skin and hair. Over the long haul of life, good eating habits will provide you with one great day after another, days full of energy, not days that are sluggish, sleepy, and suffused with brain fog.

We each have our own particular balance of meat and fish, fruits and vegetables, potatoes, whole grains, pasta, and rice that suits us. But that individualized and satisfying food profile, which I hope you'll construct for yourself, won't contradict—if it's healthy—this chapter full of basic dos and don'ts. It will only benefit from them.

Food transforms our bodies. Let's consider two aspects of how and what effect it has on you.

First, food provides the building blocks for this extraordinary, high-energy, mobile body that you live within. Clearly, a healthy spectrum of nutrients is necessary if your body is to function at its optimum level. Simply put, if you want to maximize the operation of all your metabolic systems, you need to ingest a powerful regimen of vitamins and minerals, proteins, fats, and carbohydrates, packaged in the highest-quality foods.

Such foods will also provide you with a wide-ranging panoply of antioxidants—a molecular defense team no one can do without. That certainly exemplifies the positive aspect of eating: the right foods strengthen and protect you from a myriad of glitches in your machinery, while providing the fuel for function. But keep in mind that you can in no way receive all the nutrients that are essential for optimal health from food alone (more on this in the next chapter and also in appendix C). Supplements are also very important. Recent research now provides good, scientific evidence for that statement—one I have been making to my patients for decades.

Second, of course, it's important not to eat the wrong things. There are foods—refined sugar, white flour, margarine, rancid or hydrogenated oils—that aren't much good for anyone at any time. To eat such things in significant quantities is self-destructive. In addition, there are other foods—often sternly criticized—that are fine in moderation. I'm referring to meat, fish, fowl, eggs, and butter. How much of these animal sources of food you should be eating is very much a matter of individual metabolism and preference.

Some people do best on a high protein, low carbohydrate diet. More fat and cholesterol inevitably accompany more protein, but if supplemental antioxidants are taken, I have not seen a problem with that. Dr. Robert Atkins of New York City, author of many popular weight-loss and diet books, has advocated this approach for decades. For many people, that system seems to work best. Weight becomes easier to control, energy levels increase, and sense of well-being improves. Despite what you may hear from other sources, I have not seen an increase in risk factors in my practice for atherosclerosis on that type of dietary regimen.

Increased pancreatic release of insulin with age is probably as much a risk factor for heart disease as dietary fat. Fat and cholesterol have only been shown to cause problems if they become oxidized. Supplemental antioxidants help to prevent that.

Many people, especially those who have gained weight, develop resistance to insulin as they grow older. The pancreas releases insulin into the blood in response to dietary starches, sugars, and other carbohydrates. Carbohydrates cannot be taken into cells and utilized without insulin. As cells become resistant to insulin, the pancreas must increase production. This leads to persistent high blood levels of insulin, which block the breakdown of fat, leading to weight gain and obesity. Excess insulin also contributes to high blood pressure, abnormal cholesterol and triglyceride ratios, and increases the risk for heart disease and stroke. The new buzz word for that set of conditions is "Syndrome X."

As cells become increasingly resistant to insulin, the pancreas may not be able to produce enough to control blood sugar. If sugar levels in the blood, called glucose, become too high as a result, it's diagnosed as diabetes—type II, adult onset diabetes. If the pancreas becomes totally exhausted and loses its ability to produce insulin, type I, insulin-dependent diabetes can result. It's not uncommon for adults to be somewhere between those two extremes, producing more insulin than normal, but still keeping their blood sugar within limits. Those are the people who often do better with less carbohydrate and more protein in their diet.

Conversely, some people feel better on a low-protein, low-fat, and high complex carbohydrate diet. This is the approach originally proposed by Nathan Pritikin as the Pritikin Diet and more recently popularized by Dr. Dean Ornish, called the Ornish Diet.

Dr. Barry Sears proposes a diet intermediate between these, with attention to the types of carbohydrates that require less insulin to utilize. He calls this the "Zone Diet" and stresses carbohydrates with a lower glycemic index, which trigger less insulin release by the pancreas.

Others may prefer a strict vegetarian diet and do very well on that regimen.

Any of those dietary regimens *can be* acceptable and *can be* healthy if the general guidelines in this chapter are followed.

A word of caution: you can't have it both ways! Those who follow a high protein, low carbohydrate approach, as advocated by Dr. Atkins, must be truly conscientious about carbohydrate restriction or they'll gain weight. Carbohydrates can be very addicting and difficult to give up.

Not all calories are equal in their ability to cause obesity. Some people remain obese while eating less than one thousand calories daily. Others stay slim eating fives times that amount. An important difference, I believe, is in blood insulin levels and insulin resistance at the cell membranes. The way to reduce insulin is to eat less of those carbohydrate foods that promote insulin release by the pancreas. The average American gains one pound per year between age twenty and age seventy. From age seventy on, average weight tends to decline.

I'm going to propose in this chapter an eating plan that seems to work well for the vast majority of my patients and which has substantial scientific evidence to back it up. If you follow this plan and then modify it in the direction of what makes you feel most energetic and healthy day in and day out, you should have no difficulty creating a truly formidable pro-longevity diet. And my name for that diet is:

The Anti-Free Radical, Prolongevity Diet

That's exactly what this chapter will introduce you to. Electron-hungry free radicals are byproducts of merely living.

The reason that the mammals on this planet—including humans—are capable of leading such incredibly active lives is because they use an extraordinary high-activity fuel, oxygen. Watch the way a wood fire burns: It's consuming oxygen, as well as logs, for fuel. In a similar way we also consume oxygen but in a more controlled fashion; and we pay the price in potential oxidative damage to our cells, enzymes, genes, and chromosomes.

Fortunately, our bodies have constructed a formidable network of defense against the almost explosively reactive substance that fuels it. Our bodies make every effort to stay saturated with antioxidants—substances that inhibit oxygen free radical formation and neutralize free radicals already formed, many of which are necessary in controlled amounts for life and health. Nevertheless, it has been estimated that there are enough free radicals released in our bodies during every twenty-four-hour period to subject each and every one of our cells to a thousand hits by unbalanced electron-stripping molecular fragments. This is war!

It turns out that many of the prime sources of pathologically active, potentially disease-causing free radicals can be avoided with dietary manipulation. For example, we can avoid consuming processed fats and oils (particularly the more rancid ones I'll tell you about later) and stop eating highly processed and refined foods, which have been stripped of essential protective antioxidants and other micronutrients.

This is a decided departure from the widely practiced but outmoded "no eggs, less animal fat" dietary approach to minimizing the risk of cardiovascular disease.

The anti-free radical diet establishes a five-pronged approach to this internal war against free radicals:

1. It reduces consumption of foods that metabolize most readily into excess free radicals.

185

2. It supplies optimal amounts of free radical scavenging nutrients.

3. It provides ample quantities of trace and ultratrace nutrients necessary for healthy metabolism—for healing, immunity, and manufacture of antioxidant enzymes.

4. It utilizes less-oxidizing food preparation methods to minimize lipid peroxidation before consumption.

5. It reduces intake of calorie-dense fatty foods, helping to avoid obesity and reducing dietary sources of lipid peroxide free radicals. *Lipid* is a medical word meaning "fat," and lipid peroxidation is a process whereby free radicals disrupt fats and fatty membranes within and surrounding cells in the body, a process, therefore, of cell destruction—just what you don't want your food to be doing to you. I'll frequently be mentioning peroxidation as a bad feature of those foods you don't want to eat too much of.

Unlike some rigidly restrictive nutritional regimens, this anti-free radical, prolongevity diet doesn't take all the fun out of eating. You needn't swear off prudent amounts of properly prepared eggs, butter, shrimp, steak, and other foods high in cholesterol that have long and mistakenly been maligned as lethal.

This is not really a diet in the traditional sense, inasmuch as there are no meal plans to follow. It's more of an eating, cooking, and food-selection program, designed to help you eat wisely, to take in the maximum possible protection from free radical-induced illness and nutritional deficiencies.

Here are the prolongevity guidelines that my patients and I attempt to live by:

Moderate consumption of dietary fats and oils—especially the processed or hydrogenated varieties—to a level equaling 30 to 35 percent or less of total calories consumed. Although essential fatty acids are necessary for healthy skin, arteries, blood, glands, nerves, and, indeed, all cells, most health experts agree on the wisdom of lowering total fat intake from the average American intake of 45 to 50 percent of calories. They disagree, however, on how low is low enough. Some maintain no more than 10 percent—a truly Spartan requirement—of total calories is optimal. My recommendation is to stay in the more easily managed 30 to 35 percent range, paying careful attention to the sources, the processing, and the quality of your fats and oils. I've encountered very few patients willing or self-disciplined enough to remain on a diet of only 10 percent fat calories for an extended time.

Let me digress somewhat and discuss the rationale for dietary fat restriction. It's a scientifically accepted fact that cholesterol (itself a fat-soluble, lipidlike sterol) and other dietary fats and oils only become harmful after they are damaged by oxygen and oxygen radicals. The resulting oxidized and peroxidized lipids are not only toxic in their own right but are also precursors for further chain reactions of free radical propagation.

This potential for disease-causing effects can be avoided in two ways:

1. fats and body tissues can be protected by antioxidants, or

2. fats and cholesterol can be restricted, starving the oxidative process of fuel.

The latter is the approach recommended by most nutritionists because it has been proven to protect against age-associated diseases. Unfortunately, it also deprives the body of essential fatty acid nutrients and cholesterol, which are necessary for many vital functions of the body's antioxidant defenses. And it takes much of the joy out of eating. Lipids and cholesterol are also the raw material from which a large number of hormones and vitamin D are produced within the body.

In recent years, a sizable body of scientific evidence has accumulated to show that antioxidant deficiency is much more important than an excess of fat and cholesterol as a cause of disease.

Low blood levels of vitamin E were shown in a World Health Organization study to be one hundred times more significant relative to atherosclerosis and cardiovascular disease than high blood levels of cholesterol. The daily intake of a broad spectrum multivitamin preparation, similar to that which I prescribe for my patients, was shown in a study by the University of California to increase life expectancy by as much as six years.

The multiple supplement formula I recommend as a minimum for all adults is listed in chapter 15. For older patients at risk for atherosclerosis and other diseases of aging, I recommend additional antioxidants.

To use a metaphor, if the fire in a furnace is breaking through the walls and threatening to burn the house down, it can be combated in two ways:

1. put out the fire by depriving it of fuel, which is analogous to the low-fat, low-cholesterol approach to nutrition, including the use of cholesterol-lowering drugs; or

2. fireproof the house and make the furnace walls more resistant, protecting the structure from damage while continuing to benefit from a steady supply of fuel.

The latter is a more commonsense approach and is in accordance with my recommendations.

Except for a very few people who have inherited a potentially lethal gene for exceptionally high cholesterol (sometimes 400 mg/dL or higher), it makes more sense to me to increase the intake of antioxidant nutrients, available both in food and in other nutritional supplements, rather than deprive the body of enjoyable foods that contain vital fats and cholesterol. Many recent scientific studies support that approach.

The pharmaceutical industry sells billions of dollars' worth of cholesterol-lowering drugs each year and advertises aggressively against the nutritional supplement approach. Drug company salespeople routinely call on doctors to tout the benefits of a variety of expensive prescription drugs to lower cholesterol. They rarely, if ever, inform doctors that cholesterol-lowering pharmaceuticals also have antioxidant activity and antiplatelet activity, similar to antioxidant vitamins, and that it's quite possible that reported benefits stem from those properties, rather than from reduction in blood cholesterol. But there are no patents to protect large profits that might otherwise subsidize a physician-education program on the benefits of vitamins, minerals, and supplemental antioxidants.

The untold part of the story is that reducing the quantity of fat is not nearly as crucial as eliminating the wrong kinds of fat. Contrary to popular myth, it's not the saturated (animal) fats that are the "bad guys" and the polyunsaturated fats (liquid oils of vegetable or seed origin) that are the "good guys." It's sometimes exactly the reverse—especially if the unsaturated oils have been exposed to light, heat, and air in the extraction, bottling, and food preparation process. And they almost always have been.

Saturated fats, such as we find in butter, eggs, beef, lamb, and pork, can be eaten more safely when prepared properly. The saturated fatty acids that they contain are not—under normal conditions—easily subject to the cell-damaging lipid peroxidation that I told you about a moment ago.

In sharp contrast, the polyunsaturated fats that are commercially sold (vegetable and seed oils) often undergo extensive lipid peroxidation that damages their molecular structure. This can begin the very moment those oils are extracted from the foods in which they naturally occur. Consumption of such chemically altered oils disrupts our normal metabolism, impairs cell membrane integrity, and helps to initiate a mutation process that contributes to cancer and plaque formation in the arteries.

The richer the oil in polyunsaturated fatty acids (which contain trace amounts of unbound metallic elements) and the longer it is exposed to heat, light, and atmospheric oxygen, the greater the health threat. The poorest-quality oils are customarily used in the manufacture of salad dressings and mayonnaise since their rancidity can be so easily masked by heavy seasoning.

Even the so-called cold-processed oils, premium priced in health food stores, may also, in excess, cause damage to the arteries. Just as soon as the oil is extracted from its source—the soybean, peanut, corn kernel, walnut, sesame seed, etc.—it begins to oxidize. It stores best if refrigerated and protected from bright light.

Heating vegetable oils to fry foods greatly compounds the problem. When an oil is heated, the rate of oxidation increases rapidly, doubling with every ten degrees' centigrade rise in temperature.

Hydrogenation, such as takes place during the commercial preparation of margarine, vegetable shortenings, and products like nondairy creamers and nondairy whipped toppings, converts polyunsaturated fats and oils into dangerous transfatty acids. Transfatty acids have been twisted out of shape and cause cell walls to be weakened. Most of the baked goods and so-called junk food in your local supermarket contain large quantities of hydrogenated oils, which have the commercially pleasing property of extending shelf life.

Do you protect your health by substituting margarine for butter? Hardly!

Margarine is clearly more toxic. Contrary to the image of the attractive Indian maiden on the package, by the time corn oil margarine reaches your table, it is almost completely unnatural. Not only have its original ingredients been drastically altered, but its free fatty acids have been combined with harsh chemicals and treated by petroleum-based solvents. The last defenders of this extraordinarily unattractive food began to lose their confidence in 1993, when Dr. Walter Willett and his team of researchers published another chapter in the findings of the Harvard Nurses Study, a many-decades-long project following the health fortunes of eighty-five thousand nurses. It turned out that the women who were eating the equivalent of four or more teaspoons of margarine daily had a 66 percent greater risk of developing heart disease than women whose consumption was very low or who didn't consume the synthetic butter substitute at all. As for butter itself, there was no indication of increased cardiovascular risk among the women who ate it.

There's no need to be fanatic about it. It's what you do all the time, not what you do occasionally that really matters. No one likes a fanatic! Here are some suggestions that will help you optimize intake of dietary fats to maximize both the length and the quality of your life:

• When eating beef, lamb, pork, or veal, always select the freshest, leanest meat available. Aged meats owe their enhanced flavor to rancid fat. Trim excessive visible fat before preparation. To satisfy your beef hunger, choose dishes such as casseroles or stews that provide smaller individual portions of meat than a roast or a steak. When a recipe calls for hamburger, buy the leaner variety, precook it, and drain off the fat before adding it to the dish. You may also wrap cooked hamburger in several layers of paper towels and squeeze out residual fat. Pressure cooking leaner and less expensive cuts of beef will make them quite tender.

- Substitute fish and lean poultry for other meat. Even the leanest beef you can buy has more than twice as much total fat as skinless white chicken, turkey meat, or fish. Remove the skin and visible fat before cooking or eating poultry.

- Eliminate fried foods from your diet when practical. Become adept at greaseless cooking. Use greaseless, nonstick cookware (Teflon is one popular, easily available brand), which never needs oils or fats to keep food from sticking. Slow roast meats, fish, and poultry.

- Use dairy products with a lower fat content, such as one percent fat cottage cheese and low-fat or nonfat skim milk. Cottage cheese and milk labeled as containing one percent fat by weight actually contain 10 percent of total calories as fat. Whole milk contains 40 percent of calories as fat. Be wary of imitation dairy products. Pseudo–sour cream, for example, is often made with hydrogenated oils. But nonfat and low-fat cream cheese, yogurt, and sour cream are now easily available. Read the labels carefully, however. Nonfat products often compensate for flavor loss by adding lots of sugar.

- Use more vinegar, lemon juice, garlic, onion, or herbs for salad oils and dressings, and use tomato and other fruit juices for rich sauces and gravies. One of my personal favorites for seasoning is Mrs. Dash Original Formula. Garlic and onion have the additional benefit of being rich sources of dietary antioxidants.

• Limit your intake of hidden fats, which are often hydro-genated or otherwise altered. Read labels. Be wary of pies, cakes, puddings, ice cream, and similar desserts. Nonfat ice cream is now available, but it remains very high in sugar.

In addition to the above tips on ways to moderate fat intake, select foods to maximize nutritional value. Avoid the "white plague" foods—white flour, white rice, and refined white sugar—whenever practical. It's not necessary to be extreme about it, but the less the better.

Much of the American public now suffers from a form of "overconsumption malnutrition." Their diet contains too many calories that have been stripped by the food industry of most of the trace nutrients necessary for proper assimilation and for protection from external and internal degradation. The high-speed milling of grains such as wheat, rice, and corn results in the reduction or removal of more than twenty nutri-ents, including essential fatty acids and the majority of miner-als and essential trace elements. Ideally, your refined carbohydrate intake should be carefully limited whenever you have a convenient choice—use little white-flour breads, or crackers, cereals, pastas, or snacks made from highly processed, nutrient-depleted starches. If the label says "enriched" be especially cautious. Very little of what has been removed is added back in the so-called enrichment process.

In comparison with the nutrients that naturally exist in the grain of wheat as it's growing in the field, the approximate percentage of each removed during the production of white bread is listed below:

90% of vitamin A
77% of vitamin B1
80% of vitamin B2
81% of vitamin B3

72% of vitamin B6

77% of vitamin B12

50% of pantothenic acid (B5)

86% of vitamin E

67% of folic acid

60% of calcium

40% of chromium

89% of cobalt

76% of iron

85% of magnesium

86% of manganese

71% of phosphorus

77% of potassium

16% of selenium

30% of choline, and

78% of zinc and copper!

At most, only four of these vitamins—B1, B2, B3, and iron—are put back during the so-called enrichment process. And, if a deficiency does not exist, iron supplementation has the potential to accelerate age-related free radical damage in your body.

Refined white sugar likewise lacks many vital nutrients, including those ingredients, such as chromium, needed for its metabolism. Each spoonful you consume depletes those very nutrients from body reserves needed for its utilization. Insulin cannot metabolize sugar in the absence of adequate chromium. (More on this in appendix B at the end of this book.) The average American's annual intake of sugar and high fructose corn syrup is now 137 pounds. That almost equals consumption of one's total body weight in sugar yearly.

Many of the enzymes involved in free radical protection— your body's own internally produced antioxidant team, including catalase, superoxide dismutase, and glutathione peroxidase—require the nutrients listed above that are lost in the refining process. Without these control enzymes, free radicals are generated at an ever-increasing rate.

Other molecules that neutralize unwanted free radicals include beta-carotene, vitamin E, vitamin C, the trace elements selenium, zinc, copper, manganese, and the amino acids methionine and cysteine. Without these substances—which are virtually zapped out of your foods during processing—the body cannot protect itself. The best way to cut down on your consumption of refined carbohydrates is to up your intake of natural, unrefined foods, particularly whole grain products and green and yellow vegetables, which should ideally make up about 50 percent or more of your total daily calories. Here are some tips for an unrefined, healthy diet.

• Read labels carefully. Choose sugar-free whole grain cereals and whole grain breads. Eat brown rice and whole wheat, buckwheat, or soy pasta products. Steer clear of products with the telltale "enriched" notation. When you translate "enriched" into honest English, what you get is "almost totally impoverished."

• Cook from scratch more often. There are hidden additives and refined sugars in hundreds of processed and ready-to-eat foods that we don't normally think of as sweetened, such as canned vegetables, salad dressings, catsup, biscuit mix, TV dinners, mayonnaise, and steak sauce. Hydrogenated fats and oils are also more likely to be found in factorymade, convenience, and processed foods.

• Avoid foods that contain refined sugars. These are often disguised with names like sucrose, dextrose, corn sweeteners, corn syrup, maltose, invert sugar, raw sugar, brown sugar, and turbinado, and fructose. Most so-called raw and brown sugar is nothing more than refined white sugar, colored with a little molasses or caramel to darken it. The same is true of much so-called whole wheat bread, in the form of colored white flour. (More on what constitutes good bread can be found in appendix B.)

• Be wary of sugar-free soft drinks or so-called diet drinks, especially those that are cola-flavored. They often contain phosphates, which, in excess, can disrupt calcium metabolism. Evidence is also accumulating to incriminate the artificial sweetener aspartame as a cause of many adverse affects on the body.

• Eat mainly whole foods that have not been fractionated—such as fresh fruits and vegetables, whole grains, peas, and beans—whenever possible. Eat the food whole (the fruit instead of the juice) and the entire vegetable (potatoes with skins). Train yourself to shop more around the fringes of your supermarket, the outer aisles where fresh produce, dairy products, meat, and fish are sold. Avoid the brightly colored packages of highly processed foods. It's been said, with some truth, that the cardboard boxes often contain more nutrition than the contents.

• Increase dietary fiber. Dietary fiber binds bile acids and promotes speedier movement through the digestive tract. This reduces the time they're subjected to putrefaction, oxidation, and reabsorption—which can otherwise increase production of oxidized LDL cholesterol by the liver. Increasing your consumption of dietary fiber is easily accomplished as follows:

1. Eat more root vegetables, such as potatoes, parsnips, yams, beets, carrots, turnips, etc. Other high-fiber vegetables are spinach, cabbage, sprouts, cauliflower, broccoli, and eggplant.

2. Eat whole-grain bread and pasta. There is very little fiber value to bread made of milled white flour.
3. Start the day with a high-fiber breakfast cereal. Examples include oatmeal, brown rice, whole grain cereals, and rolled wheat.
4. Add miller's bran or oat bran to your favorite recipes. It has an innocuous texture and flavor.

• Reduce salt consumption—no matter how often you've heard it, it bears repeating. Many foods, even with a prudent diet, will already have salt added, and deficiency is highly unlikely. Cut back on excessive salt. Not simply because of the link between sodium intake and high blood pressure, but also because cell walls damaged by free radicals lose some of their ability to maintain a proper sodium gradient. Excessive sodium can leak into cells that are already compromised, causing further metabolic impairment. Small capillary walls, damaged by free radicals, may leak plasma into soft tissues, causing swelling and edema. A free radical weakened sodium pump is less able to remove excess sodium from within cells. (A word of caution here: A few people feel weak on salt restriction.)

1. Lowering your salt intake is easier said than done. Food manufacturers add it to foods that rarely taste salty. For example, would you suspect it to be an ingredient in Kellogg's corn flakes? Jell-O pudding? Low-fat cottage cheese? All three have salt added, but the real surprise is how much. One serving of pudding may contain 404 mg of sodium chloride— one-third the amount in a steeped-in-brine dill pickle.
2. Since salt is a likely ingredient in any processed food not specifically labeled "No Salt Added," here are the most practical ways to cut back:
   a. Reduce voluntary salting. Use a saltshaker sparingly at the table. When cooking, substitute garlic, onion

powder, kelp powder, herbs, and natural spices in recipes that specify salt.

b. Avoid excessive high-salt condiments, soy sauce, for example, and prepared steak sauces, gravies, and relishes.

c. Moderate your intake of salt-laden foods, including smoked fish, delicatessen-style meats, canned soups, pickles, pretzels, potato chips, and similar snacks.

Learn to cook the anti-free radical way. More often than you might realize, it's not what you cook but how you cook that adds to health problems. As a general rule, the faster the food is cooked and the higher the heat it's exposed to, the more health-depleting the changes that occur. Heat speeds up the chemical reactions of peroxidation. Heat also destroys many vitamins.

Here are four rules for anti-free radical cooking. Again, there is no need to be fanatic about it, but make wiser choices when practical.

1. Limit broiling over hot coals.

If you're one of millions of suburban homeowners who relish backyard patio cooking, you won't welcome this news: Your cherished charcoal broiler is a free radical generator. Charring food oxidizes it, producing free radical precursors, and this is the reason that charbroiled and smoked foods may be carcinogenic. That sizzling steak (hamburger, hot dog, and chicken breast), salted, seasoned, and grilled to tasty perfection, becomes coated with compounds—similar to those found in tobacco tars—called aromatic polynuclear hydrocarbons. They potentially generate excessive free radicals, including singlet oxygen, against which the body has little inherent defense. In fact, the smoke from a single steak's fat drippings contains as much of the carcinogen benzopyrene as the smoke from approximately three cartons of cigarettes, i.e., six hundred coffin nails. Fortunately, optimum intake of nutritional supplements can go can long way toward protecting against this.

Grilled hamburger presents a special problem. Because of its high fat content and the large surface area exposed to air and heat, it is the most easily oxidized of all meats. In addition, iron and copper (potent free radical catalysts) are crushed out of the meat's cells and into the fat and juices during the grinding process, accelerating oxidation, making it more potentially dangerous, especially if the ground beef is a few days old. Also, hamburger is a rapid breeding ground for bacterial contamination that may have occurred in processing—a cause of the recent *E. coli* scares.

The solution is to have your hamburger fresh ground and extra lean and use it at once. Keep it well refrigerated or frozen until cooked. Steer clear of fast-food burgers. Or at least, ensure that they are well done. And, if you can't give up your backyard grill, trim all visible fat from steaks, chops, and ribs, and take both skin and fat off the chicken. Above all, take your vitamins as antioxidant protection.

2. Fry foods less often.

The oxidation of fat and oil used in frying added to the oxidation of the fat found in the food itself add up to a double whammy. Even animal fats—chicken, pork chops, fish, or eggs, normally thought of as saturated—also contain some unsaturated fatty acids and cholesterol, both of which oxidize easily.

Animal experiments have shown oxidized cholesterol to be so damaging that if as little as 1 percent of the cholesterol in your diet is consumed in its oxidized form, atherosclerosis may be triggered.

Here's what that means in practical terms.

Eggs are okay unless you fry them in hot fat. In its natural state or when a fresh egg is either soft-boiled or poached with intact yolk unexposed to air, its cholesterol content remains unoxidized and is itself an excellent free radical scavenger and nutrient. The situation is reversed when the egg is scrambled, powdered, or cooked into a recipe. Then the cholesterol is heated and exposed to oxygen and becomes partially oxidized into a number of potentially cell-damaging, toxic byproducts. The same holds true for animal meat foods, which contain preformed cholesterol—including most meats and many types of shellfish, poultry, and seafood. When fresh baked, poached, or steamed, with most visible fat removed, these are high-quality foods. Fried, they become less wholesome.

Be doubly wary of restaurant-fried foods, where highly oxidized (rancid) fat is often used over and over again in deep fat fryers, with infrequent changes.

3. Learn to cook without overcooking.

Invest in a crock-pot or a wok. Both methods rarely allow foods to exceed 212 degrees, the boiling point of water. Below that, lipid peroxidation takes place much more slowly. When using a wok, add a little water (and a minimum of oil) to prevent food from getting overly hot.

4. Avoid using aluminum cookware.

Ordinarily, aluminum cookware should not be a problem because the human body does not absorb much aluminum. However, studies indicate that aluminum does build up in the tissues of some disease victims. Aluminum deposition has been proven to occur in the arteries of atherosclerosis patients and in the brains of Alzheimer's victims and some types of Parkinson's sufferers. Therefore, although not a proven cause, it makes sense to limit your exposure as much as possible, especially since aluminum is already widely used as a food additive, is in our drinking water, our medicines, and is used in cosmetics and toiletries.

Eat an abundance of fresh whole grains and vegetables—and do so as soon as possible after purchase while they are still fresh.

While all methods of food storage result in gradual loss of nutrients, some are decidedly worse than others. Freezing delays deterioration, but major nutrient losses can still occur if the food is blanched prior to freezing. Prolonged storage results in vitamin loss and progressive oxidation, even if frozen.

For instance, asparagus left unrefrigerated for three days before use has lost most of its B-complex vitamins, perhaps before you even get it home. The same holds true, in varying degrees, for other veggies. The fresher the better. When possible, buy farm-fresh produce at farm stores and markets and only as much as you can serve and eat shortly after purchase. Or better yet, grow your own—not very practical for most people these days.

I emphasize fresh, whole foods, close to their natural state, because a wide spectrum of vitamins and micronutrients remain at healthy levels, and nothing artificial has been added. Organically grown produce is preferable although more expensive and less widely available.

You can get a high concentration of vitamin C from green peppers, broccoli, Brussels sprouts, strawberries, spinach, oranges, cabbage, grapefruit, and cauliflower. To beef up the beta-carotene content of your diet, select carrots, sweet potatoes, cantaloupe, apricots, peaches, cherries, tomatoes, and asparagus. Other fruits and vegetables rich in B-complex vitamins are peas, corn, potatoes, lima beans, and artichokes.

Here are some general rules that apply to choosing and eating fresh foods:

1. When shopping for food, give top priority to freshness.

Of the alternatives after fresh, frozen should be second choice, preferably if not blanched or thawed and refrozen.

Third place goes to dried foods. They deteriorate quite slowly, but lose most vitamin C and vitamin A content because of heat involved in the drying process. In last place come canned foods. Canning of fruits and vegetables causes the most nutrient loss—as much as 50 percent or more. Also, excessive sugar and salt are usually added in the canning process.

2. Eat your vegetables raw whenever possible.

Concoct your own low-fat salad dressings by combining whatever spices, herbs, and condiments you like with nonfat yogurt, nonfat buttermilk, or 1 percent low-fat cottage cheese. Most modern cookbooks have a variety of recipes for delicious oil-free salad dressings. When you must cook vegetables, undercook them. They should retain their crispness. Steaming is much preferred to boiling. Microwave cooking preserves nutrients better than high-heat methods, but I'm not sure how much it disrupts molecular structure and energy. I microwave when I'm in a rush, but not routinely.

3. Eat fresh foods as soon as practical after purchase.

Return frozen foods that reveal telltale refrozen signs when opened (food covered with ice crystals or a sheet of ice). Keep close track of food stored in your freezer. Check "use by" dates on containers. (Yes, they're often there, but you have to hunt for them.)

Avoid excess caffeine, soda, and alcohol—as well as chlorinated drinking water containing chemicals. Simply put, people fare better without excess caffeine. Drinking more than a few cups of coffee per day may increase your risk of ill health.

Soft drinks, especially cola-flavored varieties (sugar-free or not) skew the body's delicate calcium/phosphorus balance, already a problem since the typical American diet contains twice as much phosphorus as calcium instead of the optimal one-to-one ratio. Where does all the extra phosphorus come from? Red meats have many times more phosphorus than calcium, another good reason to rely more on poultry and fish for high-quality protein. Many carbonated beverages have phosphate buffers to prevent the carbon dioxide from forming carbonic acid. Extra phosphorus also comes from processed foods laced with preservatives, many of which are phosphate-based.

Why is the calcium/phosphorus ratio so important? When it is out of balance, excessive calcium tends to leak into cells, deposit in soft tissues, and accelerate aging. When cells are overwhelmed by calcium, they die.

One or two moderate-sized alcoholic beverages a day—no more—is a safe limit for most persons, except for pregnant women, people with seriously compromised health, and those with a predisposition for alcoholism. Alcohol metabolizes to become acetaldehyde, a potent free radical precursor and cross-linker.

Inspect your water supply. Water is no less important to our bodies than food, yet we have good reason to fear for the quality and safety of our water supply. Public water supplies often contain numerous added chemicals that are potentially harmful. Artificially softened water can also be dangerous because of its excess salt and its occasional abundance of lead and cadmium, two of the more potent toxins.

I advise that you drink well water only when you are certain the well is far away from any source of commercial toxic waste and that, if you are suspicious, you have it tested. Otherwise drink bottled, spring, or distilled water or equip your tap with a tested and effective water purifier. Reverse-osmosis water purifiers are quite good, especially in conjunction with activated charcoal. Activated charcoal filters are almost as effective by themselves and simpler to install.

When planning meals to suit this anti-free radical prolongevity diet, here are the foods I recommend you eat each day:

• Vegetables. At least two generous servings of a variety of fresh vegetables, especially those green and yellow varieties known to be the chief food sources of the important antioxidants beta-carotene, vitamin C, and vitamin B complex. There are many vital nutrients in vegetables that we have not yet identified. A list of nutritious veggies includes: artichokes, asparagus, beet greens, snap beans, lima beans, navy beans, broccoli, Brussels sprouts, cabbage, carrots, cauliflower, collards, chard, yellow corn, kale, kohlrabi, lentil, mushrooms, green peas, pumpkin, sauerkraut, snow pea pods, spinach, squash, succotash, winter squash, sweet potatoes, yams, tomatoes, turnip greens, turnips, water chestnuts, and zucchini. There are many more.

• Salads. Eat two servings a day from any combination of raw vegetables. You can add extra anti-free radical potency by the liberal use of chicory, Chinese cabbage, cucumber, endives, escarole, lettuce, parsley, pimento, chives, red and green peppers, dandelion greens, watercress, radishes, scallions, garlic, onions, and leeks. Season salads with lemon juice, herbs, or oil-free or low-fat dressings. Olive oil on salads is less likely to be oxidized because it is a monounsaturate.

• Fruits. Two to three small to moderately sized fresh servings a day, not canned, cooked or juiced, with choice depending on individual taste and what is seasonally available. Good fruit sources of antioxidants are apricots, bananas, cantaloupe, all melons, oranges, tangerines, papayas, peaches, plums, prunes, lemons, limes, pineapples, tomatoes, black currants, raspberries, rhubarb, and strawberries.

• Protein Foods. Two servings a day consisting of the following: four- to six-ounce portions of lean beef, pork, veal, or lamb. Intersperse red meat with chicken, turkey, or fish. Use water-packed canned tuna.

- Eggs. One or two a day, preferably fresh-boiled or poached, with yolks intact, not scrambled, fried, or cooked into other dishes.

- Cereals and breadstuffs. Eat two to four servings a day of unrefined whole-grain products. These include such breakfast cereals as oatmeal, all bran, and shredded wheat, pasta, breads, muffins, or crackers made from stone-ground whole-grain flours, and brown rice. Sparingly spread breads with butter, if desired, never margarine.

- Dairy products. Drink mainly low-fat or nonfat skim milk. Use low-fat or nonfat cheeses and other dairy products. Soymilk is a nutritious substitute.

- Beverages. Decaffeinated coffee is fine (no more than three or four cups a day of nondecaffeinated coffee). Drink no more than three cups of milk per day. Limit to a maximum of two ounces of alcoholic liquor, eight ounces of wine, or two twelve-ounce cans of beer. Limit most soft drinks. There is no limit on naturally carbonated spring water.

- Desserts and snacks. Choose fresh fruit, dried fruit, puddings, sherbets, and gelatins made at home from sugar-free recipes. For an occasional treat, have a handful of freshly cracked, unsalted nuts; fresh-roasted chestnuts; dried raisins; or a slice of fat-free, homemade sponge or angel food cake. Be aware that many dried fruits are soaked in sugar solution before they are dried. Avoid aspartame—Equal or NutraSweet—whenever possible. If you must use an artificial sweetener, in my opinion saccharin (Sweet'N Low) is the lesser of the evils.

Eating the anti-free radical prolongevity way is not difficult and can easily be made very pleasurable. Learn to shop, cook, and eat so that your food intake prepares you for a hearty old age. It can be an exciting adventure. You have much to gain—a longer, healthier, happier life.

Finally, as you'll see in the next chapter, it's simply not possible to receive optimum amounts of vitamins, minerals, trace elements, and antioxidants from diet alone. The regular use of nutritional supplements can do much to minimize potential free radical problems.

15

LIFE AFTER CHELATION—EIGHT THINGS YOU CAN DO TO LIVE HEALTHIER, LONGER

How much longer do you want to live?

Ten, twenty, fifty years? Forever?

How healthy do you want to be?

Not such frivolous questions once you realize healthful life extension is within your control regardless of your current health status.

Not too long ago there was scant evidence that you could forestall age-related diseases, or reverse those already in progress. Now we know better. More than 80 percent of atherosclerotic disease is largely self-inflicted, preventable, and partially reversible. The "killer diseases"—cancer included—are not always the inevitable consequences of age, genetic background, or environmental exposure as once thought.

You can live better and longer simply by improving lifestyle habits that subvert health and shorten life. The time has come to stop asking what your doctor can do for you (to paraphrase J. F. Kennedy) and ask instead what you should be doing for yourself.

Here are eight things you can do to maximize your longevity. Whether you adopt one, or all, the change for the better will move you closer to able-bodied health.

1. Quit smoking. Don't use tobacco in any form!

Only a modern-day Rip Van Winkle, asleep for the past five decades, hasn't heard that smoking is a disease-producing, life-shortening habit. Even with all the warnings, the full hazards are not well known. It's not just the tar and nicotine content of cigarettes that will do in the smoker, but so long a list of poisons and toxic reactions, were they put on the package, there'd be little room left for the brand name. If you're still unconvinced, here is the newest damning evidence.

• Smoking greatly increases internal free radical production.

• A pack-a-day smoker can absorb as much ionizing radiation in one year as he would were he to have two hundred chest x-rays, according to University of Massachusetts doctors who have found radioactive isotopes, particularly polonium-210 (from cheap phosphate fertilizers used in growing tobacco) and also radioactive isotopes of lead, "highly concentrated" on smoke particles. Radioactivity causes tissue damage by producing excess free radicals.

• Tobacco smoke contains tars, free radical-producing polynuclear aromatic hydrocarbons that can cause genetic mutations, leading to both cancer and atherosclerotic plaque. Free radical agents are absorbed from tobacco through the lining of the mouth, even when the smoker does not inhale. Chewing tobacco and snuff are even worse.

• Cigarette smoke contains acetaldehyde, a hazardous chemical cross-linker that causes undesirable chemical bonds between large molecules via free radical reactions. Abnormal cross-linking results in wrinkled skin and inelastic, hardened arteries. A form of aldehyde is also used in embalming fluid. In that regard, tobacco may begin the embalming process, even before death.

- Cigarette smoke contains toxic heavy metals—lead, cadmium, and arsenic—that contribute to free radical pathology by limiting antioxidant enzyme activity and also poison metabolism directly in other ways.

- Cigarette smoke contains carbon monoxide, which reduces the blood's oxygen-carrying capabilities, and nitrous oxides, which increase free radical production.

- Tobacco also triggers allergic reactions and increases risk of sudden death from heart attack. A team of scientists at Cornell University Medical College found that smokers and nonsmokers alike manifested allergic skin reactions to tobacco glycoprotein, an antigen found in tobacco leaves and cigarette smoke. Allergic reactions can disturb heart rhythms, weaken heart contractions, and interrupt coronary artery blood flow.

- Tobacco contains nicotine, which causes blood vessels to constrict. Nicotine increases blood pressure and reduces blood flow to the coronary arteries and other vital organs.

- "Pot" smokers are no better off since they suffer the additional hazard of smoking paraquat-treated cannabis (marijuana). Paraquat toxicity speeds the cross-linking process.

My clinical experience adds another dimension. In almost every case where chelation therapy failed to improve a patient's health, he or she had continued smoking. In some instances, smokers improved for a while, but their conditions worsened more rapidly, compared to nonsmokers, once treatments stopped. I don't refuse to chelate smokers, but I warn them in advance that the benefit will be less and will not last as long.

Smokers can reverse damage by giving up their habit. In one California study of heart attack victims, most of those who ceased smoking had less evidence of arterial plaque within eighteen months to two years after the day they stopped. A Swedish study of men who survived a first heart attack, comparing "quitters" with "nonquitters," revealed that those who quit had nearly half the death rate of those who continued to smoke. Other studies came to similar conclusions. The more time that elapses after your last smoke, the closer your coronary heart disease risk comes to that enjoyed by people who have never smoked. After approximately ten years, the comparative risks seem to equalize.

2. Avoid smoke-filled environments.

"Pass-along" smoke is also deadly. Inhaling secondhand smoke makes the heart beat faster, increases blood pressure, and raises blood carbon monoxide levels. There can be almost as much cadmium, tar, and nicotine in the smoke drifting from a smoldering butt as that inhaled by the puffer.

Those who insist on their right to smoke close to you are infringing on your right to life. A recent Environmental Protection Agency study revealed that the innocent nonsmoking bystander suffers damage equivalent to smoking four or more cigarettes per day. Children of smoking parents suffer twice as many childhood ills—particularly respiratory ailments—as children of nonsmokers. If you live or work side by side with a smoker, as far as your lungs, heart, and circulatory system are concerned, you are a smoker, too. A study of nonsmoking Japanese wives revealed that their health correlated with whether they lived with smoking or nonsmoking husbands. Twice as much illness and lung cancer turned up in women married to smokers.

What to do?

• Encourage your employer, bridge club, and professional associations to adopt and enforce the "no-smoking" rule, or segregate smokers in another room.

• When making hotel, restaurant, or airline reservations, specify the nonsmoking section.

• If you cannot convince your mate to quit (or if you entertain smokers), equip your home with room air purifiers.

• When in a captive situation, position yourself upwind of the smokers.

3. Get regular exercise.

The most exercise many Americans get each day is slamming the car door after parking as close as possible to the supermarket. Their motto: "Never take a step when you can 'let your fingers do the walking.'"

A growing subculture of exercise buffs ascribes to the opposite philosophy: "Don't walk when you can run." Some have developed jogger's addiction, the compulsion to run five or more miles every day, come what may, including hurricanes, drought, pestilence, or alien invasion.

To what extent does exercise help the heart? There's conflicting evidence as to whether anything less than extremely vigorous physical workouts provide significant protection against heart attacks. Several studies suggest an athlete must "sweat" during exercise to substantially reduce the risk of myocardial infarction.

There is no argument that habitual, lifelong exercise is a plus. It improves the quality of life in many ways. It helps burn off fat, suppresses appetite, aids circulation, reduces anxiety and depression, and increases general fitness.

Anyone past their mid-thirties who has been only moderately active, however, would be wise to go slow before indulging in activities as strenuous as marathon running. At least one study has shown that half of all sudden deaths occurred immediately after or during severe or moderate exercise. According to Dr. Meyer Friedman (of "Type A" fame), once someone has suffered a heart attack, he should refrain from overly strenuous exercise: he should not jog, run, play handball, racquetball, or tennis.

What if you've always been a desk-bound exercise phobic who's hated the very thought of doing anything strenuous? Is it too late to start?

When a group of formerly sedentary men and women aged sixty and older were enrolled in an exercise program, it took only seven weeks for them to acquire fitness levels equal to that of average forty- to fifty-year-olds, according to Canadian researchers.

The best exercise? For anyone who feels up to it—or who works up to it—I recommend "wogging" (a term coined by Dr. Thomas W. Patrick Jr. of Fort Lee, New Jersey). As the name suggests, wogging is the happy compromise between walking (not enough exertion) and jogging (too strenuous and joint-jarring).

To get the most benefit from wogging (which Dr. Patrick defines as "walking fast for pleasure, exercise, and physical fitness, at different rates from brisk to rapid"), start off easy, about a block or two at first, and then work up to thirty minutes or more per day. This minimum amount of moderate exercise will help get more oxygen to more cells and that's a boost to the body's free radical defenses.

Various "stair-climbing" exercise machines, rowing machines, and exercise bicycles offer an excellent indoor form of exercise. Thirty to forty-five minutes of aerobic exercise several times each week will do much toward maintaining cardiovascular fitness.

Almost as important as regular exercise is avoiding inactivity. Sitting down and relaxing while watching TV may be your favorite pastime, but two researchers have documented cases of blood clots brought about by long sessions of TV-induced immobility. Their suggestion? Get up and walk around the TV set every twenty minutes or so. The same good advice pertains to sedentary workers, whose heart attack rate has been found to be twice as high as employees in comparable, but more physically demanding, jobs.

When asked what he does to keep in shape, the director of the Cardiovascular Center at New York Hospital-Cornell Medical Center, Dr. John Laragh, reported that he believes in informal, nonstructured exercise that keeps the muscles loose, lively, and active.

"You can do that around any office if you stay on your feet and move around," he said. "I rarely sit at a desk unless I'm writing. I use the stairs instead of the elevator whenever I can."

Exercise needn't be tedious. As Dr. Laragh said, you're not aiming at becoming a gladiator. Get fit for the fun of it, choosing relatively simple and pleasant ways to stay limbered up. You'll be more motivated to continue on a regular basis if your fitness regimen provides real pleasure as you keep your body in motion.

Among the things you might do to up your activity level: get a dog; go dancing; join a bowling team, health spa, aerobics class, or swim club; visit museums and art galleries; stand up and walk around while you talk on the phone; get rid of your remote TV tuner; do housework to music; park the car and walk; volunteer to tend a toddler once a week; team up with a hiking buddy; walk—don't ride—the golf course; disconnect telephone extensions at home.

4. Limit alcohol consumption.

To avoid health hazards, limit alcohol consumption to one ounce of pure ethanol per twenty-four hours (at most, that's two eight-ounce glasses of beer, two small glasses of wine, or two small shot glasses of hard liquor). A relatively healthy adult can normally detoxify that amount of alcohol without exceeding the free radical control threshold.

Indeed, research conducted by the National Heart, Lung, and Blood Institute suggests that up to one ounce of alcohol per day may be good for your heart. Their long-term nationwide study involving thousands of adults confirmed a positive relation between limited alcohol consumption and favorable HDL cholesterol levels. Red wine also contains antioxidants that protect against free radicals.

Drink more than that, however, and you soon head for trouble. Alcohol is converted to acetaldehyde in the liver (closely associated with formaldehyde, or embalming fluid), which auto-oxidizes in the body in the presence of unsaturated lipids, damaging cells membranes in an explosion of destructive free radicals. Our bodies contain sufficient special enzymes to metabolize the acetaldehyde, but this internal detoxification mechanism can be overwhelmed if too much alcohol is consumed at one time, or too fast, or if long-term, heavy consumption results in free radical damage to the liver's enzyme production. Once acetaldehyde is left free to roam, it creates more and more dangerous free radicals, further hampering the body's built-in detoxification machinery, setting up a destructive chain reaction. That's how cirrhosis of the liver can occur.

Avoid alcohol entirely if you are chronically ill or suffer serious degenerative disease.

5. Take a scientifically balanced nutritional supplement daily.

In the proper combination, supplemental nutrients will reinforce the body's natural antioxidant and free radical scavenging abilities, especially when supplements are taken with meals. Mealtime is when the digestive process can trigger increased levels of free radical production. The supplement you select should protect against nutritional deficiencies and at the same time provide an adequate and balanced spectrum of the many different vitamins, trace elements, and other nutrients necessary for antiaging benefit and protection against undesired free radical damage. It's simply not possible to get the full, optimum doses of these nutrients from food alone. To give you an idea of what to look for, six tablets a day of the product I recommend to patients, taken three tablets at a time, twice daily directly after the morning and evening meals, provides the following:

NUTRITIONAL SUPPLEMENT FORMULA

	Six tablets daily contain
Vitamin A (from fish liver oil)	10,000 IU
Beta-carotene (natural, D. salina)	15,000 IU
Vitamin D-3 (from fish liver oil)	400 IU
Vitamin E (d-alpha tocopheryl succinate)	400 IU
Vitamin K-1 (phytonadione)	60 mcg
Vitamin C (L-ascorbate, hypoallergenic)	1,200 mg
Vitamin B-1 (thiamine)	100 mg
Vitamin B-2 (riboflavin)	50 mg
Niacin (vitamin B-3)	50 mg
Niacinamide (vitamin B-3)	150 mg
Pantothenic acid (d-calcium pantothenate)	400 mg
Vitamin B-6 (pyridoxine)	50 mg
Folic acid	800 mcg
Vitamin B-12 (an ion exchange resin)	100 mcg
Biotin	300 mcg

Choline (bitartrate)	150 mg
Calcium (citrate, ascorbate)	500 mg
Magnesium (aspartate-ascorbate, chelate)	500 mg
Potassium (aspartate-ascorbate)	99 mg
Copper (amino acid chelate)	2 mg
Manganese (amino acid chelate)	20 mg
Zinc (amino acid chelate)	20 mg
Iodine (kelp)	150 mcg
Chromium (ChromeMate(r))	200 mcg
Selenium (amino acid complex)	200 mcg
Molybdenum (amino acid chelate)	150 mcg
Vanadyl Sulfate	200 mcg
Boron (aspartate-citrate)	2 mg
PABA (para-amino benzoic acid)	50 mg
Inositol	50 mg
Citrus Bioflavonoids	100 mg

(Contains no yeast, corn, wheat, sugar, or other sweeteners, artificial colors, flavors, or preservatives. Sealed with lot number and expiration date printed on the label.)

This formula was originally devised by me more than twenty years ago, with the help of a team of nutritional scientists, including biochemists knowledgeable about the complicated interrelationships among vitamins, minerals, and trace elements in metabolic functioning. The combination of ingredients has been upgraded and modified from time to time to reflect the latest research.

This formula provides the foundation. It contains a physiologic balance of a wide spectrum of micronutrients and will prevent imbalances when other things are added. Many other supplements can be added to this foundation with age. Most of my patients over age forty also take co-enzyme Q10, extra vitamins C and E, and a variety of other items, depending on symptoms.

While proper nutritional supplementation is good, more is not necessarily better. Taken to excess, substances such as iron, aside from being toxic, may speed free radical damage by catalyzing lipid peroxidation. The very elements essential for health and life are also potential health destroyers. Iron supplements should be taken only with evidence of a proven deficiency and for only long enough to replenish body stores.

The person most apt to be at risk of over supplementation is the true believer—the dedicated "vitamintologist" who avidly follows "panacea-of-the-month" revelations in a variety of magazine articles. Tinkering with nutritional supplementation is dangerous. Dietary micronutrients only work well in harmony with each other. A deficiency in one or an excess of another may disrupt metabolism. For example, if you take high doses of zinc without copper, you may create a copper deficiency; take zinc without selenium, and you may increase the risk of cancer. You'll find more on that in appendix B.

Trace element supplementation, not a task for an amateur, should take place under the supervision of a trained nutritionist or knowledgeable health practitioner, who can tailor recommendations to individual needs based on dietary history, medical evaluation, and biochemical tests.

6. Learn to relax.

Do you plunge full speed ahead in a determined, forceful, never-enough-time manner?

Are you hard-driving, super-ambitious, goal-oriented?

Is your conversation full of "shoulds," "musts," and "ought-tos"?

When there's a tough job to be done, are you the first to sign up for it?

While the connection between personality and coronary artery disease is not fully established, there does seem to be a certain type of individual (first described by Dr. Meyer Friedman and Dr. Ray Rosenman as "Type A's") that is unusually prone to heart attacks.

The typical Type A, according to researchers, is an intensely competitive overachiever, never satisfied with current accomplishments, always in hot and heavy pursuit of a more elusive goal. So-called Type A's, in the most extreme cases, are also aggressive, hostile, impatient, easily provoked, impulsive, argumentative, and bossy. As a class, those who fall within this category release more adrenaline and have a greater rise in blood pressure when stressed.

Heart disease also correlates with anger. The more anger one harbors, the more heart disease will occur. This anger can be hidden and directed inward—often manifested as depression.

Type A's are not just high-salaried executives holding down heavy responsibilities in high-pressure jobs. One study found 15 percent of stay-at-home housewives were Type A's and they, too, had twice the rate of coronary heart disease that Type B women had.

Animal researchers have discovered there are Type A bunnies. In an experiment, fourteen rabbits were divided into two groups based on the way they reacted to minor stress. When fed an atherogenic diet, the "high stress" rabbits were found to develop "significantly greater atherosclerotic plaque than those rabbits that exhibited low stress," according to a report issued from the University of Southern California School of Medicine. They also had slightly higher blood pressure. In a yet more recent study, primates fed low-cholesterol, low-fat diets were divided into two groups—one group stressed, the other not. The healthy diet protected the unstressed animals from atherosclerosis, but not the stressed group.

If you're a Type A, chances are you know it. Friends, relatives, your children, and your spouse are always urging you to relax and slow down.

Should you recognize yourself as being a Type A person, what can you do about it? Basic personality cannot easily be changed, but behavior can be modified. Behavioral change experts suggest trying the following techniques:

• Practice some form of relaxation. If you have no particular liking for any structured relaxation technique, try twenty minutes a day in a hot bath, letting your mind drift in pleasant daydreams. If you're religious, use prayer. You might consider a course in meditation.

• Rid yourself of unnecessary and unproductive activities. Delegate responsibilities. Learn to say "No!" to requests you would prefer to turn down. Pretend you have only six months left to live and set new work/play priorities.

• Free yourself from self-imposed time traps.

If you suffer from the "hurry sickness," with never enough time to get things done, chances are your problems are self-created. To break the time bind, you might stop wearing a watch, avoid committing yourself to closely timed appointments, and do not schedule back-to-back activities. Develop the *mañana* philosophy. When things pile up, tell yourself "Tomorrow is another day."

A corollary to anger and relaxation is attitude in general. A recent Mayo Clinic study shows that optimists live longer than pessimists and suffer less disease.

7. Expand and cherish your social network.

Science has begun to prove what was long suspected: The health of your heart depends to a large extent on the state of your emotions. While the body/mind connection has never been in question, only lately has it come to light that a good defense against a heart attack might come from maintaining strong, close, satisfying personal relationships.

One of the most startling—and convincing—illustrations can be found in the tale of Roseto, Pennsylvania, famous as the "town without heart attacks."

The story might never have come to light had not a local doctor, who practiced there for seventeen years, told several colleagues he had never seen heart disease in anyone under the age of fifty-five. Temple University researchers found Roseto was indeed unique, with a death rate from myocardial infarction less than half that of the neighboring towns and half that of the rest of the country.

Scientists moved in to find out what was happening and thus began a nineteen-year study. For the first five summers, health professionals set up clinics and examined most of the over-twenty-five Roseto population, recording histories, blood studies, urinalyses, blood pressures, and electrocardiograms. To their surprise, they discovered the Rosetoans—mostly Italian-Americans—were almost identical to people in neighboring communities. Many were overweight, ate fatty diets, smoked, drank, rarely exercised, and had an equal amount of the bad habits normally considered precursors to coronary artery disease.

Medical sociologists dug in, determined to find an answer. Finally, they came up with the one striking feature about life in Roseto which set it apart. In their published report, the scientists said: "The people of Roseto adhered to a tenaciously held lifestyle, which reflected Old World values and customs. It was characterized by predictability and stability."

According to the researchers, Roseto was a town where everyone knew everyone else. It had a clannish quality. Family relationships were extremely close and mutually supportive. Neighbors helped neighbors; the elder generations were cherished and respected, and retained their authority throughout their lives; men and women had well-defined roles, and men were the uncontested heads of their families; personal problems were worked out within the family or with the help of the local priest; social life revolved around the family, village celebrations, and religious festivals—until the early 1960s, when things began to change.

Some young people, who went away to college, never returned. The first generation Rosetoans began to die off. The birth rate declined. Church attendance went down. Interdenominational marriages increased. Rivalries cropped up. The more affluent began to show off their wealth. Competition increased. Men joined the country club; women slimmed down and dolled up. Fast-food drive-ins attracted the diners who formerly congregated at the five family-owned restaurants. The old, close, clannish feeling was gone, and at the same time there began a striking increase in death rates from heart attack. Within ten years, Roseto's heart attack rate was identical to that of neighboring communities. What went wrong?

The doctors who followed Roseto's history believe it was the strong family ties, sense of community, and supportive social networks that once protected residents against coronary disease. As those bonds weakened, so did defenses against heart attack.

While we cannot re-create the Roseto that once was, we can recognize the value of love and friendship to our health and our lives. When Israeli doctors conducted a long-term study, they found a far lower incidence of heart attacks among happily married men who considered their wives loving, loyal, and supportive.

Loving and being loved is so crucial, it even extends to man's relationship with the animal kingdom. Psychologists studying the bond between people and their pets have discovered that men who had heart attacks, and owned a dog, were less likely to have a second heart attack than their petless counterparts. Physiologists have documented beneficial physiological reactions when man and animal interact in a loving manner; pet your dog or cuddle your cat and your blood pressure goes down and your heart rate decelerates.

The implications are clear. Tender loving care must take its place along with the more traditional heart disease preventive measures.

8. Switch to a health-promoting diet.

We've left the best for last. The development of atherosclerosis takes decades. If the diet that contributes to this disease is changed to one that treats and prevents it, damage can be repaired and health restored. To help you achieve that goal, switch to my "Anti-Free Radical Diet," as described in chapter 14.

16

WHAT ABOUT BYPASS SURGERY AND ANGIOPLASTY?

In the two centuries since a classical description of angina pectoris as a symptom of coronary heart disease first appeared in the medical literature, there has been little for physicians to offer patients in the way of effective treatment. Only recently have effective therapies become available.

The coronary arteries provide blood and energy to the muscle of the heart—the pump. That pump must contract seventy times per minute or so to propel blood throughout the body. It cannot stop to rest, like other muscles, or death will occur in minutes. If the flow of blood is partially restricted by atherosclerotic plaque, the heart continues to pump, but it hurts, similar to a leg cramp. That type of pain is called angina pectoris, which is Latin for "chest pain"—commonly shortened to angina.

Angina is often triggered by stress or exertion because at those times the heart must work harder to pump more blood. The discomfort can also be dull, manifesting as a pressure sensation. It can radiate to the arms, neck, back, or upper abdomen and may even mimic indigestion. Angina is usually relieved by rest. If the blockage is severe, a part of the heart muscle may die, causing a so-called heart attack—in medical terms, a myocardial infarction.

Right through the end of the 1940s, nitroglycerin was the standard treatment for this type of pain. At the onset of an angina episode, a victim placed a tiny white pill under the tongue and waited for the uncomfortably tight, strangling sensation and pain to subside. But because the condition underlying angina is usually progressive, pain became more frequent and severe, and less amenable to nitroglycerin relief.

The seriously afflicted had little choice but to learn to live (or die) with their condition. Since exertion predictably triggered frightening discomfort, many were forced to adopt curtailed lifestyles, giving up former work and play activities.

In 1950 a seemingly miraculous remedy captured attention. Surgeons developed a new operation called internal mammary artery ligation, which involved surgically tying off the mammary artery, which carries blood to the exterior chest wall. Because this artery is located near the heart, surgeons hoped this action would force more blood to flow through other arteries in the vicinity (including coronary arteries) and ease the pain of angina.

Results exceeded the most optimistic expectations. Remarkably, up to 90 percent of patients reported either total pain relief or dramatic symptom improvement. The operation, hailed as a miraculous advance, was widely advocated by many members of the medical profession. Enthusiasm mounted; angina victims lined up; surgeons maintained three-month waiting lists. The operation's effectiveness went unquestioned and untested, for almost ten years.

But then, as now, there were skeptics within the medical community. There was too much enthusiasm to suit some discerning physicians who doubted that the procedure deserved such universal acclaim, inasmuch as it had a dubious scientific rationale. The doubters arranged to verify the surgery's effectiveness with a research protocol, which would be unacceptable under today's more rigid ethical standards.

They set out to test the procedure by dividing surgical candidates into two groups, each equally afflicted with angina. All subjects were told they were to undergo ligation surgery and went through identical hospital protocols with only one important difference: One group did have the ligation operation whereas the control group was taken into the operating room, anesthetized, and then subjected to a sham operation. Their chests were opened, then closed. When they awoke, they were told their operations had been successful.

To the astonishment of the entire medical community, the surgeons included, both groups reported relief from pain of angina and increased tolerance to exercise. But, the group that had undergone the sham surgery fared better than those who had undergone the genuine operation. It was the first time medical researchers proved that the placebo effect extends to surgery; and, not surprisingly, when word got out, the number of operations plummeted.

Unlike the well-documented time sequence of placebo effect, patients don't experience full benefit from EDTA chelation therapy until three months after therapy is completed. And benefit continues for many months or years thereafter, even without further therapy. This result is very unlike any placebo effects ever reported. Placebo effects occur at once and last only a few months at most. At the time when placebo effect would fade is when relief from chelation therapy reaches its peak. I have never seen reports of placebo effect that lasted as long as six months.

What has this to do with current methods of treating angina?

More than one leading scientist has expressed the belief that, in many cases, coronary artery bypass graft (CABG) surgery, one of the most common major operations performed in the United States today, is the current equivalent of the sham surgery of the 1950s. Said one, "My own suspicion is that a placebo might do just as well, and not cost $50 thousand, the usual price tag of a coronary bypass operation." According to the American Heart Association, in 1995, 1,460,000 angiograms were performed at an average cost of $10,880 per procedure. This resulted in 573,000 bypass surgeries at an average cost of $44,820 and 419,000 percutaneous transluminal (balloon) coronary angioplasties (PTCAs) at an average of $20,370 each. The total bill in 1995 was $50 billion, or $137 million per day—$5.7 million per hour. That's big business! The total annual cost of cardiovascular disease in the United States, including medications and disability, is approximately $274 billion per year.

Despite the commotion surrounding the bypass procedure, it has never been conclusively proven to do much more than relieve the pain of angina (except for approximately 15 percent of patients who meet very specific selection criteria). As with any symptom-relieving treatment, there is a real possibility that the placebo effect is at least in part responsible.

The scientific references listed at the end of this chapter were used to gather source material for the following discussion of bypass surgery.

Heart bypass may also serve as a type of "surgical beta blocker," with an action paralleling that of a group of drugs that diminish pain by interfering with nerve impulses, which can trigger arterial spasm in the coronary arteries and increase heart muscle contraction, resulting in angina. Although not widely known, it is impossible to perform the operation without partially disrupting the nerves that stimulate the beta receptors on arteries and heart muscle. Nerves that transmit the pain of angina are also transected.

Is bypass surgery, like the operation that preceded it by some twenty years, undeservedly popular?

When the Office of Technology Assessment was commissioned by the United States Congress to review the case for surgery for coronary artery disease, it was not greatly impressed. A panel of government consultants, which included leading academicians from the nation's most prestigious medical schools, reported to Congress: "For more than half a century, surgeons have believed that an efficacious surgical approach to coronary artery disease is possible. Prior to the modern bypass operation, five different operations were developed and advocated enthusiastically. Although all five operations were ultimately abandoned as of no value, initially they were alleged to be efficacious, with reports in the medical literature claiming 'objective' evidence of benefits."

Noting that "coronary bypass surgery seems to give excellent symptomatic relief from angina pectoris . . . but the improvement diminishes with time," the government panel of experts cautioned that there was a historical lesson to be heeded, pointing out that "the possible placebo effect (of bypass surgery) needs to be kept in mind because: the initial results are similar to previous operations; nonsurgical treatment also produces good results; and the methods of evaluation of symptomatic relief are experiential."

The chief of cardiology at the Montreal Heart Institute, Dr. Lucien Campeau, is a cardiovascular specialist who suspects that long-term relief of angina pain results from what he calls a "pain-denial placebo effect." Dr. Campeau came to this conclusion after studying 235 patients angiographically three years after their coronary artery bypass operations, discovering that even in cases where grafts had reclosed, patients unexpectedly reported being improved or angina-free.

A report in the *Journal of the American Medical Association* (JAMA) once again documented angina pain relief in 75 percent of patients who have bypass surgery. Shortly thereafter, an article in the *New England Journal of Medicine* stated 75 percent of angina patients' pain is also relieved with nonsurgical therapy. In effect, these two highly authoritative articles are saying bypass surgery works no better than noninvasive therapies.

Many patients who opt for this operation have a real need to believe in its effectiveness. They have a huge emotional as well as financial investment in a successful outcome, often having been scared into believing that this surgery is the only way to save their lives.

Claims that the operation prolongs life to any significant degree are still being debated. When the Harvard University School of Public Health put coronary bypass surgery to the test, they concluded it is often unnecessary. The Harvard study involved 142 men who had all "flunked" a treadmill exercise test and had other evidence of extensive coronary atherosclerosis. Each had been advised to undergo the bypass operation.

But when this group of surgical candidates was referred to Harvard specialists for a second opinion, surgery was rejected in favor of medication, diet, and exercise. After keeping tabs on these 142 men for anywhere from twenty months to twelve years, the Harvard researchers found their death rate exactly what would have been expected had the men been operated on (provided, the study pointed out, they survived the operation, which has an operative mortality of 2 to 3 percent).

Contrary to the claims of cardiovascular surgeons, bypass surgery does little to improve the outlook for survival, according to the Harvard report.

A study by Dr. Wilbert Aranow at the University of California comparing atherosclerotic heart patients treated surgically with those treated medically revealed no evidence of increased survival or lowered heart attack risk. Nor did studies conducted by Duke University Medical Center find reason to suggest that coronary surgery prolongs life when compared with medical management.

An analysis of 1,101 consecutive patients with coronary artery disease was made by the Division of Cardiology at Duke—490 had surgery; 611 were treated nonsurgically. At the end of four years, there was no significant difference in the survival rate between the surgically treated and the medically treated: survival was 82 percent for the first group, 78 percent for the second.

In a study reported in the *New England Journal of Medicine* by Paulin et al., 686 patients with stable angina were followed for twenty-two years. Of that number, 322 received bypass surgery and 312 were treated medically. Although there was an early survival benefit in high-risk patients (up to a decade), long-term survival rates thereafter became comparable in both treatment groups. At twenty-two years, the cumulative survival rate was 25 percent in the medically treated group but only 20 percent in patients who had received bypass surgery.

Brain damage is a common complication of bypass. In another article published in the *New England Journal of Medicine*, Roach et al. reported on mental impairment following bypass surgery: "Adverse cerebral outcomes after coronary bypass surgery are relatively common and serious; they are associated with substantial increases in mortality, length of hospitalization, and use of intermediate- or long-term care facilities." Five years after bypass, 23 percent of patients showed an abnormal mental decline in ability to make sense of spatial relationships, and an additional 16 percent had persistent impairment in their ability to remember words. Six percent of bypass patients suffered more serious brain injury, including dementia, stupor, stroke, and epileptic seizures.

The Newark *Star-Ledger* reported in March 1999 that the statewide New Jersey death rate as a complication of bypass surgery was 3.37 percent and that in some hospitals it was as high as 8 percent.

One important study of long-term results following bypass was the Coronary Artery Surgery Study (CASS), in which 780 patients were followed for more than twelve years.

When that long-awaited ten-year, government-funded study was released, it offered little encouragement for advocates of cardiovascular surgery. The study was conducted at eleven prominent medical centers: the University of Alabama, Alabama Medical College, Boston University, the Marshfield (Wisconsin) Clinic, Massachusetts General Hospital, Milwaukee Veterans Hospital, New York University, St. Louis University, Stanford University, Yale University, and the Montreal Heart Institute. Seven hundred and eighty volunteer patients with coronary heart disease were divided into two groups. Half had bypass operations; the other half had non-surgical treatment consisting of prescription drugs and advice to start exercising sensibly and avoid risks like smoking, overeating, and consuming too much fat in their diets.

After many years of follow-up, results now show that the most severely diseased 15 percent of patients who submit to bypass surgery actually did get a measurable benefit. But even for those few, the death rate was higher during the first two years following surgery because of surgical complications.

The 15 percent of patients who did get small but statistically significant benefit from bypass fell into the following three categories: (1) high-grade obstructions of the left main coronary artery system, including the left anterior descending artery, without adequate collateral flow around those blockages; (2) high grade blockages of all three major coronary arteries without adequate collateral flow; and, (3) greatly reduced pumping action of the heart. Patients who met one or a combination of those criteria eventually experienced a small increase in survival rate lasting a few years. In that small group, from the second to fifth years, the death rate was about 10 percent less compared to patients who do not have surgery.

That advantage was lost from the fifth to the tenth year and, indeed, from the tenth year on.

In other words, 15 percent of patients have a 10 percent lowering of death rate five years after surgery, which amounts to a 1.5 percent overall lowering in death rate following bypass surgery, as it is now performed. Remember, from 2 percent to 8 percent (depending on the hospital) die immediately as a complication of surgery, and surgery therefore results in an increase in death rate during the first two postoperative years. I wonder how many patients would agree to undergo bypass if those statistics were clearly presented to them in advance?

It is, nevertheless, true that some patients—those who have been carefully selected and who suffer severely impaired quality of life from coronary heart disease—do experience dramatic improvements following either surgery or balloon angioplasty. I'm not opposed to those procedures. I refer patients when I think they need that kind of therapy. But for patients whose condition is stable and not worsening at a dangerous rate, I'm definitely opposed to immediately and aggressively resorting to invasive, expensive, and potentially fatal procedures without first trying treatments that involve less risk and much lower cost.

A very significant finding of the CASS study is the fact that the death rate for patients who did not have surgery or angioplasty was only 2 percent per year. That is quite a low death rate for patients with serious heart disease, and seems hardly to justify the risk of death or other complications from surgery or invasive procedures.

After six years, 92 percent of surgical patients and 90 percent of the medical patients were still alive. The researchers concluded that tens of thousands of bypass operations every year were unnecessary and could be eliminated. That's fine as far as it goes. But the real question to be asked is, did the scientists speak out boldly enough?

Many think not. There is good reason to suspect they were extremely conservative in their estimate of the annual number of unneeded operations and downplayed their statements concerning the percentage of bypasses that could safely be avoided.

As Dr. Eugene Braunwald, professor of cardiology at Harvard Medical School, pointed out in the *New England Journal of Medicine*, the data were already obsolete when the CASS study came out, inasmuch as it was collected before the advent of newer calcium-channel blockers and improved beta-blockers. "Non-surgical therapy has not stood still during the last six years," Dr. Braunwald noted, challenging the validity of findings that exclude recent advances in nonsurgical cardiovascular treatment.

Were the researchers too kind to proponents of surgery? If, in the majority of cases, bypass surgery is no better than less drastic treatments, does it do any harm?

There is no question that bypass surgery and angioplasty can often relieve symptoms of angina and are suitable for patients whose quality of life is greatly impaired by coronary heart disease not relieved by medicine. They must be willing to accept the risk of greater than a 2 percent chance of death and the 25 percent incidence of other serious complications from those invasive procedures.

A U.S. government pamphlet entitled "Medicine for the Layman—Heart Attacks," published by an agency of the U. S. Government, noted clinical investigations have yet to determine whether bypass surgery improves or impairs heart function. As stated by this booklet, "There is no evidence yet that bypass surgery makes the heart pump better—some evidence exists that bypass surgery may actually decrease efficiency."

Medical authorities are increasingly critical of bypass surgery. Thomas A. Preston, M.D., professor of medicine at the University of Washington School of Medicine and chief of cardiology at the Pacific Medical Center, Seattle, Washington, wrote about coronary artery bypass surgery, "As it is now practiced, its net effect on the patient's health is probably negative. The operation does not cure patients, it is scandalously overused, and its high cost drains resources from other areas of need." He further says, "A decade of scientific study has shown that, except in certain well-defined situations, bypass surgery does not save lives or even prevent heart attacks. Among patients who suffer from coronary artery disease, those who are treated without surgery enjoy the same survival rates as those who undergo open-heart surgery. Yet many American physicians continue to prescribe surgery immediately upon the appearance of angina or chest pain."

A Veterans Administration Cooperative study was also published in the *New England Journal of Medicine*. That study included 486 victims of atherosclerotic heart disease of the most critical kind with unstable angina pectoris. Half were subjected to bypass surgery, and the other half were treated without surgery. The overall results were very similar to the CASS study.

Both studies, however, were conducted prior to the use of calcium channel blockers, although beta-blockers were administered to half of the CASS patients. Both types of prescription medicines have been shown to reduce the incidence of heart attacks, decrease death rate in heart disease, and relieve angina without surgery. It is therefore not possible, without further research comparing bypass surgery with present-day medicines (including EDTA chelation therapy), to conclude whether patients would not do equally as well or even better without surgery.

An interesting report in the *New England Journal of Medicine* showed that coronary blood vessels increase in size as blockages occur. When a plaque grows to approach 50 percent of the inside diameter of a coronary artery, the artery simultaneously enlarges to compensate. The diseased artery may therefore continue to allow almost the same flow of blood as a healthy artery.

Only when plaque blockage exceeds 50 percent, and then only with strenuous exercise, does blood flow decrease enough to cause symptoms. At that point, collateral branches will often grow in from nearby arteries to maintain an adequate supply of blood, even if a major vessel becomes totally blocked.

With 75 percent blockage from atherosclerotic plaque, compensatory enlargement can cause total overall blood flow to remain equal to that in a healthy artery with only a 50 per-cent blockage. Furthermore, animal experiments show that substantially more than 50 percent blockage of a normal coronary artery is necessary to decrease heart function, even under maximum physical stress. More than 75 percent blockage of a healthy artery, without time to compensate or form collaterals, is needed to reduce heart function at rest. Nonetheless, bypass surgery is aggressively recommended in many instances with plaque blockage of 75 percent, despite adequate coronary blood flow.

An editorial in that same issue of the *New England Journal of Medicine* stated, "Those . . . who perform coronary arteriography have made one serious mistake. It consists of the unfortunate adoption of a grading system for stenoses expressed as a percentage of the arterial lumen that is compromised. This grading system implies a degree of accuracy that coronary angiography cannot achieve." It is not possible to accurately predict the three-dimensional flow of blood in an artery from two-dimensional x-ray shadows. That editorial goes on to point out that 75 percent blockage of a diseased coronary vessel is often necessary to compromise the heart under maximum physical exertion and that considerably more than a 75 percent plaque blockage is necessary to reduce function without physical exertion.

Conclusions in that report stated, "The preservation of a nearly normal lumen cross-sectional area, despite the presence of a large plaque, should be taken into account in evaluating atherosclerotic disease with the use of coronary angiography." That recommendation is often ignored at medical centers, which seem to have become dependent on financial income generated by bypass surgery for survival. The AMA has published in its official journal (*JAMA*) that 44 percent of all bypass surgery in the United States is done for inappropriate reasons.

Arterial spasm can cause anginal pain and heart attack, even without atherosclerotic plaque; and spasm is properly treated without surgery. Reversible spasm can also be triggered by irritation from the injected dye and reduced oxygen transport during angiograms, which can closely mimic blockage by plaque. Arteries are encircled by bands of muscle, like a belt around the waist. If that muscle contracts in spasm, like a belt tightening, blood flow is cut off.

Why then are patients so often told that they must have bypass surgery because arteriograms show 75 percent blockage of an artery, with no consideration for heart function, collateral branches, or total blood flow? Overall cardiac efficiency and blood flowing past and around a blockage can be measured with isotope imaging prior to recommending surgery. Noninvasive imaging of the heart using radioisotopes will often show adequate pumping action and coronary blood flow, despite extensive plaque on the arteriograms. Is it possible that isotope studies are not routinely done because they do not show the surgeon where to operate and because surgery might be canceled if blood flow were thus shown to be adequate?

Arteriograms are a major marketing tool for bypass surgery and balloon angioplasty (and now sometimes for laser vaporization or plaque removal by rotating blades). Results of catheterization and arteriograms can frighten patients into accepting unnecessary, dangerous, and expensive surgery or angioplasty, when nonsurgical treatment might be equally as effective or even more so, with much less danger and expense. The risk of harm or death to patients from the preliminary catheterization and arteriograms, although small, is still significant. I believe that arteriograms should be resorted to only when a decision is made to consider surgery or angioplasty, based on severity of symptoms and lack of response to nonsurgical treatments, including chelation therapy.

Another reason to delay surgery, whenever possible, is a recent report of accelerated atherosclerosis in arteries after they have been subjected to bypass. Plaques grow faster in bypassed arteries after surgery.

When an artery is bypassed beyond a point of high-grade obstruction, a region of back-flow and stagnant flow is created between that partial obstruction and the site of the implanted bypass. Clotting and total blockage of the original obstruction up to the point of bypass can then more easily occur, causing total dependence on the thin-walled and weaker vein graft. If that vein graft fails, the patient becomes worse off than before surgery.

One risk of surgery is the very real possibility of suffering a heart attack while still on the operating table. A number of reports suggest that that happens to as many as 3 percent of all patients and more than 10 percent of some high-risk patients, depending on the surgeon and medical center. In rare instances, the heart may refuse to resume beating when taken off the bypass machinery.

Not to be overlooked is the psychological trauma. It would be difficult to find anyone who is not terrorized by the operation. Bypass patients must also face the possibility that one operation won't do it. Reports indicate that 15 to 30 percent of vein grafts become occluded within one year of surgery.

Angioplasty has an even worse reclosure rate. As many as 50 percent of coronary arteries forced open by balloon angioplasty close up again within one year. The use of stents may improve those odds, but long-term follow-up studies have not been completed.

The ultimate damage, death.

While few deny that a bypass operation involves serious hazards, there is enormous disagreement on mortality rates, reported at anywhere from 1 to 42 percent, depending on where the procedure is done, who performs the surgery, on which group of patients, and on how data are collected. The National Heart and Lung Institute has reported the risk of death following coronary artery bypass surgery to be between 1 and 4 percent in the best of circumstances and 10 to 15 percent in the worst.

Surgical candidates are understandably quoted the most optimistic view, even though their chances of survival depend to a large degree on their age, general health status, degree of disease, and the skill and experience of the surgeon and surgical team.

The testing procedures upon which surgical decisions are based are also open to criticism. Each new diagnostic device that comes along is tacked on the ever-growing checklist. Physicians may become so captivated with space-age diagnostics that they sometimes fail to remember they're treating patients, not tests. Coronary arteriograms, electrocardiograms, radionuclide studies, nuclear ventriculograms, thallium scans, digital subtraction arteriography, ultrasound imaging, treadmill stress tests, echocardiography, ultrafast CT scans, EBCT, and PET scans can all be useful, but are overused, according to no less an authority than Dr. George Burch, professor of cardiology at Tulane University School of Medicine.

As Dr. Burch points out, "It has yet to be demonstrated

that the new information, expensively gotten, will change the way we treat patients."

What he failed to mention is how often such diagnostic procedures merely serve as an excuse to speed a patient into surgery.

The Harvard University report, previously mentioned, specifically challenged the overreliance of many heart specialists on exercise tests. The researchers noted that stress tests suggesting clogged arteries are an insufficient basis by themselves for the decision to undertake such procedures as coronary angiography as a prelude to surgery, the common current practice.

Exercise stress tests are not only inconclusive, but also carry some small risk. A study of 170,000 such tests revealed that for every ten thousand persons tested, one may die and two or three may require hospitalization. Occasionally, emergency treatment is needed. While the risk of death is very low, 0.01 to 0.04 percent, that number still seems to me significant enough to avoid indiscriminate use, considering, in many cases, that test results may be vague or misleading.

For almost thirty years, the coronary angiogram has been the diagnostic tool most revered by vascular surgeons, the one they invariably rely on for evidence of need for surgery.

In principle, the angiogram (also called an arteriogram) provides a filmed visualization of dye injected into the arteries, enabling skilled radiologists to pinpoint the precise location and extent of blockages (expressed in percentages). In actuality, that's not what happens.

Have patients gone to surgery on the basis of misinterpreted arteriograms?

"Without question," according to Dr. Arthur Selzer, cardiopulmonary lab chief at San Francisco's Presbyterian Hospital, who told a reporter he had "always been skeptical about angiographic readings, especially when expressed in percentages. That implies the evaluator is measuring something when he's just giving a visual impression of an obstruction." Radiological readings are rarely challenged. If the angiographer reports a 75 percent occlusion of the so-called "time-bomb artery" (the left main coronary, or its major branch, the left anterior descending artery), the necessity for a bypass is considered confirmed.

It was not until the National Heart, Lung, and Blood Institute (NHLBI) undertook an investigation of angiogram reliability that cardiologists were given hard evidence that coronary angiography is more art than science.

The NHLBI report, presented at an American Heart Association meeting in Anaheim, California, revealed that inaccurate assessments of arteriograms are commonplace and that when experienced radiologists evaluate the same arteriograms, they have conflicting opinions almost half the time.

The NHLBI conducted a three-pronged probe. In one study, three arteriographers, working independently, examined films of twenty-eight patients who had died within forty days of cardiac catheterizations. When their readings of the amount of occlusion of that all-important left main artery were compared with actual autopsy findings, it turned out they were more often wrong than right. In a whopping 82 percent of their judgments, the degree of narrowing was significantly under- or overestimated.

In the second stage of the research project, thirty films with distinct pathology were circulated among radiologists at three first-rate medical centers to discover how often first, second, and third opinions might agree. The discouraging results: only 61 percent of the time did two or more of the three groups reach the same conclusion.

Finally, in the third study, three months later, the same thirty films were recirculated to the same experienced radiologists, who did not know, of course, they were being asked to reevaluate films they had seen before. This time, the radiologists not only disagreed with each other, they also disagreed with themselves! In 32 percent of the readings, their second evaluations differed from their first.

One conclusion made from that study is that angiograms are, at most, accurate only to within 25 percent of the actual degree of arterial closure.

Exploding the myth of angiogram reliability has "profound implications for the diagnosis and treatment of coronary disease," declared Dr. Harvey G. Kemp Jr., cardiology chief at St. Luke's Medical Center in New York, who directed one segment of that research. Especially, he noted, since the evaluations had been conducted under the most favorable circumstances. "We had some of the best people reading the best quality angiograms available," he pointed out.

And how did the cardiovascular community respond to research that clearly indicated patients were being scheduled for surgery based on erroneous diagnoses? They didn't. Nothing's changed.

Despite findings to the contrary, the coronary angiogram remains the "gold standard" of cardiovascular diagnosis and is still considered the final word when it comes to determining if bypass surgery is indicated. Angiograms continue to be performed daily, by the hundreds of thousands every year.

To refer to the angiograms—which costs about $3,500 (about as much as a full course of chelation) and sometimes requires hospitalization—as a diagnostic test is in itself misleading when, in fact, it is an operation to get the patient ready for an operation. The recommendation for surgery seems often to be a foregone conclusion.

Occluded arteries are to be expected. Remember, atherosclerotic plaque begins accumulating before the third decade of life, and many men and women who are symptom-free and considered healthy have been found to have 75 percent or more arterial blockage when autopsied after accidental death from causes unrelated to arterial disease.

Of all the diagnostic procedures, the angiogram is often the one patients fear most. They are awake during the procedure—"Worse than the surgery which followed," some report—increasingly so now that balloon angioplasty and placement of stents are commonly performed at the time of the initial angiogram. It can be an uncomfortable procedure, involving threading a long catheter through a large puncture in an artery in the arm or groin, which is then threaded up into the heart. Dye is injected through the catheter directly into the patient's coronary arteries. X-ray films of the dye flow through blood vessels ostensibly show the location, pattern, and extent of blockages, but as we've already learned, error-ridden readings of those films degrade their accuracy and limit their diagnostic value.

It's customary for a cardiologist to get a patient's permission to proceed at once with balloon angioplasty and placement of synthetic mesh stents within arteries at the time of the angiogram, with no wait for the patient to recover and participate in that decision. Angioplasty is itself almost as risky as bypass surgery and can require emergency bypass if complications occur during the procedure. Recent data show that patients whose conditions are stable after a myocardial infarction (MI) and who are nonetheless treated with angiography and invasive procedures have a 71 percent higher mortality rate at hospital discharge, a 60 percent increase in death rate thirty days after discharge, and a 30 percent increased death rate at forty-four months' follow-up, compared to MI patients treated conservatively.

There are other risks associated with angiography. It can trigger a heart attack or stroke, either immediately or several months later, and result in torn arteries, infection, or allergic reaction to the dye. Plaques can be disrupted by the catheter, releasing small pieces, called plaque emboli, which flow downstream to block smaller blood vessels.

Finally, angiograms too often lead to a hazardous operation. Once the cardiologist requests an angiogram, the patient is frequently on the final lap of the surgical track. Angiograms can be very useful and they do have their place, but they act as such good marketing tools for subsequent surgery or angioplasty that they are utilized excessively, in my opinion.

If bypass surgery is an expensive, high-risk, limited-benefit procedure, as research indicates, why then does it continue to be the uncontested winner of the "Most Popular Operation of the Year" award? Why do almost one million Americans each year submit to surgery and other invasive coronary artery procedures, costing as much as fifty thousand dollars, which will not cure their underlying disease and has a chance of making them worse? Good question.

Bypass surgery, a dramatic operation with lots of pizzazz, has been the beneficiary of considerable media "hype." In the early 1970s, it represented the ultimate in sophisticated medical technology, made possible by newly perfected heart-lung bypass machinery. Newspapers, magazines, and TV, always eager to sensationalize science with "soap opera" appeal, zoomed in to capture every heart-throbbing (pun intended) moment of what was hailed as a medical marvel.

The general public responded as might be expected. People with angina and other heart-related problems began seeking out cardiac surgeons, sometimes without even consulting their family physician. The medical community reacted just as naively. It's not just the average man in the street who learns what's new in medicine from the television news and other news media. Surveys have shown many doctors also rely on lay publications to be informed of current medical issues. Unperturbed by the lack of proven advantages over other therapies, cardiovascular specialists embraced the new technology with questionable enthusiasm.

Almost overnight, bypass surgery became a medical fad. Indeed, in certain social circles, the sternum-splitting scar is a status symbol. An experimental procedure when first introduced, balloon angioplasty and stents soon followed. They have since become the treatments of choice for almost a million Americans each year.

"What, you haven't had your bypass yet?" one executive asks another in the locker room. The intimation is clear: only an administrator unworthy of having a key to the executive washroom would have escaped the inevitable consequences of being dedicated to one's job. More recently, the question has changed somewhat, from "Have you had your bypass?" to "How many arteries?" In several large metropolitan cities, being scheduled for cardiovascular surgery opens the door to the local "Zipper Club."

A prestigious procedure? Of course. It has a glamorous image since so many really important, famous people have had it—former Secretaries of State Henry Kissinger and Alexander Haig, King Khalid of Saudi Arabia, comedian Danny Kaye, late-night talk show host David Letterman, Larry King of CNN's *Larry King Live*, and top country music singer Marty Robbins. This two-time Grammy winner, by the way, had two bypasses: first a triple bypass, then an eight-hour quadruple bypass twelve years later. He died one week after the second operation.

The latest wrinkle among the elite is to have bypass surgery preventively. I'm not sure what that means, but when a forty-nine-year-old governor of Kentucky, John Y. Brown Jr., suffered chest pains while barbecuing the family's dinner, he was rushed to King's Daughters Hospital and twenty-four hours later underwent a triple bypass. His doctors told the press the operation was "preventive," emphasizing the governor had not had a heart attack, giving the unfounded impression the surgery would certainly ward one off.

All of which serves to prove that the rich and famous often get no better medical advice than the less privileged.

Would coronary bypass surgery have proliferated so rapidly and enjoyed such unwarranted popularity if it weren't so enormously profitable? Many critics believe it is a procedure that has gotten out of hand, primarily because of the big bucks involved.

"Every time a surgeon does a heart bypass, he takes home a new sports car," quipped one cynic, referring to the fifteen thousand dollars or more surgeon's fee that has provided some cardiovascular surgeons with incomes of one million dollars per year and more.

Nor are surgeons the only beneficiaries. Coronary artery bypass surgery and balloon angioplasty are now an estimated $50 billion a year industry, providing a financial windfall to hospitals, drug and equipment manufacturers, and guaranteed employment to a small army of highly specialized, highly paid surgical and postsurgical coronary care teams.

With medical insurance companies picking up a large part of the tab, "some nonsurgical measures may be overlooked in the rush to get cases into the operating room," according to the executive director of Maryland's Health Services Cost Review Commission. "Less expensive treatments would get greater play if patients were uninsured and had to form 'first' opinions about their own money, instead of spending someone else's," he added.

Balloon angioplasty was introduced in the early 1980s as a way to avoid costly and dangerous bypass surgery. Instead, the number of bypass operations has increased from 200,000 in 1984 to 573,000 in 1995, at a time when angioplasty procedures increased from 46,000 to 419,000 per year. Angioplasties often fail in less than a year, leading to repeated angioplasties or bypass surgery. Six percent of all angioplasty procedures require emergency surgical interventions because of complications.

In a study reported in the medical journal *Lancet* in 1997, 1,018 patients were randomized into two groups. One group received percutaneous transluminal coronary balloon angioplasty (PTCA), and the other group was treated medically. These patients were then followed for 2.7 years. The study revealed that only those patients with the most severe angina had improved pain relief; it also showed that improvement was often lost beginning a few months after PTCA, with no improvement at two years, presumably from reblockage, when compared to the medically treated group. Death and nonfatal myocardial infarction occurred in 6.3 percent of PTCA patients, compared with only 3.3 percent of medically treated patients. There were one death and seven nonfatal myocardial infarctions at the time of PTCA.

A three-year follow-up of this same group of patients was reported in the March 15, 2000, issue *Journal of the American College of Cardiology*. Patients in the balloon angioplasty group with the most severe angina had significantly greater improvements in physical functioning, vitality, and general health at both three months and one year, but not at three years. Those conclusions were related to breathlessness, angina grade, and treadmill exercise time.

Lange and Hillis reported in an editorial in a 1998 issue of the *New England Journal of Medicine* that after reviewing four recent large, prospective, randomized studies comparing invasive, aggressive therapy with conservative, noninvasive, medical management of acute coronary syndromes (angina, ischemia, and infarctions): "Studies show that routine angioplasty and revascularization [bypass] do not reduce the incidence of nonfatal myocardial infarctions or death." They go on to state that despite the fact that adverse events are similar or even greater in patients managed aggressively, physicians in the U.S. continue to choose the more aggressive and invasive approaches. Angioplasty and bypass are performed less than half as frequently on similar patients in Canada, although the incidence of myocardial infarction and death in three years of follow-up was similar. In their editorial the authors ask, "Why are coronary angiography and revascularization [bypass and angioplasty] often performed in patients with acute coronary syndromes in the United States, even without an obvious indication?"

Will criticism from within or without the medical community stem the flood to the surgical suites?

"Not likely," said one of San Francisco's leading cardiologists. "There's too much money involved. It's become a self-perpetuating industry."

Perhaps the surgeons have gotten carried away, but that's no reason for patients to play along.

Should you be advised to submit to bypass surgery or angioplasty before other treatments are fairly tried, or even considered, ask this question first: "What are my other alternatives?"

References

Anderson HV, Cannon CP, Stopne PH, et al. "One Year Results of the Thrombolysis in Myocardial Infarction (TIMI) IIIB Clinical Trial: A Randomized Comparison of Tissue-Type Plasminogen Activator versus Placebo and Early Invasive versus Early Conservative Strategies in Unstable Angina and Non-Q Wave Myocardial Infarction." *J Am Coll Cardiol* 26 (1995): 1643-50.

Boden WE, O'Rourke RA, Crawford MH, et al. "Outcomes in Patients with Acute Non-Q Myocardial Infarction Randomly Assigned to an Invasive as Compared with a Conservative Management Strategy." *N Engl J Med* 338 (1998): 1785-92.

Cashin LW, Sanmarco ME, Nessirn SA, Blankenhorn DH, et al. "Accelerated Progression of Atherosclerosis in Coronary Vessels with Minimal Lesions that Are Bypassed." *N Engl J Med* 13, no. 311 (1984): 824-28.

CASS (coronary artery surgery study) Principal Investigators and Their Associates. "A Randomized Trial of Coronary Artery Bypass Surgery." *Circulation* 68, no. 5 (1983): 951-60.

CASS (coronary artery surgery study) Principal Investigators and Their Associates. "Myocardial Infarction and Mortality in the Coronary Artery Surgery Study Randomized Trial." *N Engl J Med* 310, no. 12 (1984): 750-58.

"Coronary Angioplasty versus Medical Therapy for Angina: The Second Randomized Intervention Treatment of Angina (RITA-2) Trial." *Lancet* 16, no. 350 (9076)(August 1997): 461-68.

Glagov S, Weisenberg E, Zarins CK, et al. "Compensatory Enlargement of Human Atherosclerotic Coronary Arteries." *N Engl J Med* 316, no. 22 (1987): 1371-75.

Henderson RA, Pocock SJ, Sharp SJ, Nanchahal K, Sculpher MJ, Buxton MJ, Hampton JR, et al. "Long-Term Results of RITA-1 Trial: Clinical and Cost Comparisons of Coronary Angioplasty and Coronary-Artery Bypass Grafting. Randomized Intervention Treatment of Angina." *Lancet* 352 (9138) (October 31, 1998): 1419-25.

Lange R, Hillis DL. "Use and Overuse of Angiography and Revascularization for Acute Coronary Syndromes." *N Engl J Med* 338, no. 25 (1998): 1838-39.

Luchi RJ, Scott SM, Deupree RH, et al. "Comparison of Medical and Surgical Treatment for Unstable Angina Pectoris." *N Engl J Med* 316, no. 16 (1987): 977-84.

Peduzzi PA, Kamina A, Detre K. "Twenty-Two Year Follow-Up in the VA Cooperative Study of Coronary Artery Bypass Surgery for Angina." *Am J Cardiol* 81, no. 12 (June 15, 1998): 1393-99.

Paulin S. "Assessing the Severity of Coronary Lesions with Angiography." *N Engl J Med* 316, no. 22 (1987): 1405-7.

Popcock SJ, Henderson RA, Clayton T, et al. "Quality of Life after Coronary Angioplasty or Continued Medical Treatment for

Angina: Three-Year Follow-Up in the RITA-2 Trial-Randomized Intervention Treatment of Angina." *J Am Coll Cardiol* 35, no. 4 (March 15, 2000): 907-14.

Preston TA. "Marketing an Operation: Coronary Artery Bypass Surgery." *J Holistic Med* 7, no. 1 (1985): 8-15.

Roach GW, Kanchuger M, Mangano CM, et al. "Adverse Cerebral Outcomes after Coronary Bypass Surgery: Multicenter Study of Perioperative Ischemia Research Group and the Ischemia Research and Education Foundation Investigators." *N Engl J Med* 335, no. 25 (December 19, 1996): 1857-63.

Steinberg D, Parthasarthy S, Carew TE, et al. "Beyond Cholesterol: Modifications of Low-Density Lipoprotein that Increase Its Atherogenicity." *N Engl J Med* 320, no. 14 (1989): 915-24.

SWIFT (Should We Intervene Following Thrombolysis?) Trial Study Group. "SWIFT Trial of Delayed Elective Intervention versus Conservative Treatment after Thrombolysis with Anistreplase in Acute Myocardial Infarction." *BMJ* 302 (1991): 555-60.

Terrin ML, Williams DO, Kleinman NS, et al. "Two- and Three-Year Results of the Thrombolysis in Myocardial Infarction (TIMI) Phase II Clinical Trial." *J Am Coll Cardiol* 22 (1993):1763-72.

TIMI Study Group. "Comparison of Invasive and Conservative Strategies after Treatment with Intravenous Tissue Plasminogen Activator in Acute Myocardial Infarction: Results of the Thrombolysis in Myocardial Infarction (TIMI) Phase II Trial." *N Engl J Med* 320 (1989): 618-27.

U.S. Congress, Office of Technology Assessment. *Assessing the Efficacy and Safety of Medical Technologies* [publication no. 052-003-00593-0] (Washington, D.C.: U.S. Government Printing Office, 1978.)

Whitaker J. *Health and Healing,* no. 8, 9 (September 1998).

Williams DO, Braunwald E, Thompson B, Sharaf BL, et al. "Results of Percutaneous Transluminal Coronary Angioplasty in Unstable Angina and Non-Q-Wave Myocardial Infarction:

Observations from the TIMI IIIB Trial." *Circulation* 94 (1996): 2749-55.

Winslow CM, Kosecoff JB, Chassin M, et al. "The Appropriateness of Performing Coronary Artery Bypass Surgery." *JAMA* 260 (1988): 505-9.

17

THE FINAL WORD—TAKE
THIS TO YOUR DOCTOR

Having read this far, chances are you know more about chelation than anyone in your neighborhood: sad to say, even more than your doctor.

No problem, until the day comes when you, or someone you care for, experiences the symptoms of cardiovascular distress (angina pains, breathing difficulties, a heart attack, or threatened stroke). Where do you go? To your doctor, naturally.

The truth is, you cannot chelate yourself. Nor can you routinely turn to a chelation doctor for all of your ongoing medical management. Your postchelation health status may require the continuing surveillance of a cardiology specialist, internist, or primary care physician. Whatever treatment you choose, you will feel more comfortable with your doctor's approval and professional support.

No easy matter when chelation therapy is the issue. No matter how strong your present conviction that it has more to offer than other treatments, it is hard to stand your ground when your cardiologist's response to your question "What about chelation?" is "Never heard of it" or "Thumbs down."

If you're like most patients, you'll feel trapped and cornered. You'll feel ill equipped to contest your physician's superior medical expertise, and fearful of antagonizing the specialist who may hold your very life in his hands. It takes a special kind of courage to tell your doctor you favor an alternative that he doesn't approve of and that has not yet become widely accepted by most physicians.

Some of my patients have resolved this dilemma by keeping their chelation treatments a secret. Others have changed doctors, more than once.

There may be a better way. Let's hope that your physician is a fair and open-minded professional, who understands the importance of keeping abreast of new developments in his field. Give him the opportunity to read for himself the latest scientific treatise (completely annotated and cross-referenced to over 299 articles from the world's scientific literature) providing the scientific rationale for EDTA chelation therapy.

Take this chapter to your doctor.

If your physician is interested in learning more about EDTA chelation therapy, recommend that he or she get a copy of the 550-page medical textbook entitled *A Textbook on EDTA Chelation Therapy, Second Edition*, edited by me, Elmer M. Cranton, M.D., with a foreword by Dr. Linus Pauling. You might even purchase a copy yourself and make it a present for your doctor.

This new edition of the textbook, originally published in 1989, was completely revised and updated in 2001. It contains many important research studies with statistically significant data to support the effectiveness of chelation therapy and includes the approved protocol for safe and effective use in clinical practice.

The textbook may be purchased for $75.00 from: Hampton Roads Publishing Company, 1125 Stoney Ridge Road, Charlottesville, VA 22902, Phone (434) 296-2772. FAX (434) 296-5096. e-mail: hrpc@hrpub.com.

Elmer M. Cranton, M.D.
Mount Rainier Clinic
503 First Street South
Suite 1, P.O. Box 5100
Yelm, WA 98597-5100
Phone: (360) 458-1061
FAX: (360) 458-1661
e-mail:
drcranton@drcranton.com
website:
www.drcranton.com

Elmer M. Cranton, M.D.,
and Eduardo Castro, M.D.
Mount Rogers Clinic
799 Ripshin Road
P.O. Box 44
Trout Dale, VA 24378-0044
Phone: (540) 677-3631
FAX: (540) 677-3843

The following portion of this chapter is an actual chapter taken from *A Textbook on EDTA Chelation Therapy, Second Edition*, edited by Elmer M. Cranton, M.D., Copyright © 2001 Elmer M. Cranton, M.D.

SCIENTIFIC RATIONALE FOR EDTA CHELATION THERAPY IN TREATMENT OF ATHEROSCLEROSIS AND DISEASES OF AGING

Elmer M. Cranton, M.D. and James P. Frackelton, M.D.*

ABSTRACT: The widely accepted Free-Radical Theory of Aging has given us a coherent and unifying scientific explanation for the many diverse benefits resulting from EDTA chelation therapy and nutritional supplementation. The Free Radical Theory will be discussed foremost in this chapter. The emerging Cell-Senescence Model of Aging, however, combined

*Elmer M. Cranton, M.D., is past president and fellow, American College for Advancement in Medicine, and Charter Fellow, American Academy of Family Physicians. He practices chelation therapy, hyperbaric oxygen therapy, and anti-aging medicine in Yelm, Washington. James P. Frackelton, M.D., is fellow and instructor in Chelation Therapy, American College for Advancement in Medicine, and past chairman, Department of Family Practice, Fairview General Hospital, Cleveland, Ohio. He practices chelation therapy and anti-aging medicine in Cleveland, Ohio.

with the concept of apoptosis (programmed cell death), adds to the free radical explanation, and gives us a broader scientific rationale for EDTA chelation therapy plus nutritional supplementation. To be valid, the proposed mechanism of action must explain why benefit requires several months to occur and improvement continues long after therapy is completed. The wide range of reported benefits can be explained, at least in part, by reactivation of enzymes that are dependent on metal ions for function and removal of metals that act as catalysts of free radical proliferation. Suboptimal nutriture weakens trace element-dependent antioxidant enzymes. The use of supplemental micronutrients and antioxidants is an integral part of this therapy. EDTA chelation therapy combined with nutritional supplements and modification of health-destroying habits act in unison to prevent and to partially reverse many common age-associated diseases.

Introduction

The use of chelation therapy with intravenous ethylene diamine tetraacetate (EDTA) for the treatment of atherosclerosis is increasing rapidly worldwide. This practice, which began more than four decades ago, accelerates each year. Many published studies confirm safety and effectiveness of intravenous EDTA for treatment of occlusive atherosclerotic arterial disease and age-related degenerative diseases.[1-89] A very important basis for the scientific rationale of this therapy is thus the fact that it has been proven effective over and over again in clinical practice. More than one million patients have received more than twenty million infusions with no serious adverse effects—when administered following the approved protocol described in Chapter 32. Many years ago, reports of kidney damage and other adverse effects resulted from excessive doses of EDTA, infused too rapidly (greater than 50 mg/Kg/day, infused more rapidly than 16.6 mg/min).

Excessive dose-rates of infusion, especially in the presence of preexisting kidney disease or heavy metal toxicity, were responsible for occasional reports of nephrotoxicity.[64-74] No such adverse side effects have been reported when the currently approved protocol has been correctly followed.[72,73,74]

Research with laboratory animals provides further support for the effectiveness of EDTA chelation therapy.[77-83]

There has never been a valid scientific study of EDTA chelation that did not show effectiveness, although there have been reports in which positive data were erroneously interpreted as negative. Reports of negative or adverse results from EDTA chelation following the currently approved protocol have been either editorial comments and letters to the editor written by opponents of this therapy or seriously flawed attempts to discredit chelation with biased and unscientific interpretation of data, often by cardiovascular surgeons who freely admit their bias.[75,84-89]

In the last ten years, a small cluster of studies has sprouted up in the medical literature purporting to demonstrate that EDTA chelation is not effective in treatment of cardiovascular disease. Although flawed and imperfect, those studies in actuality provide only positive support for chelation. Their negative conclusions are not supported by the data.

The Danish Study

The most controversial and oft cited study of that type was done in Denmark. It was the handiwork of a group of Danish cardiovascular surgeons, with admitted bias against chelation. Results of that study were published in two medical journals, the *Journal of Internal Medicine* and the *American Journal of Surgery*. The surgeons' adverse conclusions were also widely publicized in the news media.[84,85]

The surgeons followed 153 patients suffering from intermittent claudication. The patients had such severely compromised circulation in their lower extremities that walking a city block or less would cause them to stop with pain. An endpoint measured for this study was their maximal walking distance (MWD)—the very longest distance that they could walk before pain of claudication brought them to a halt. Patients were equally divided into an EDTA group and a placebo group. In the pre-treatment phase, the EDTA group averaged walking 119 meters before pain stopped them; the placebo group was less limited at the outset and averaged 157 meters.

Treatment was either 20 intravenous infusions of disodium EDTA or 20 infusions of a simple salt solution, depending on their group. Although the study was purportedly double-blinded (neither patients nor researchers were supposed to know who received placebo and who received EDTA), the researchers later admitted that they broke the code well before the post-treatment final evaluation.

Both groups showed improvement, and the investigators concluded that the improvement was not statistically significant. This Danish study turned many people against chelation; but, in rather short order, the integrity of the study was called into question. It was learned that the researchers had violated their own double-blind protocol. Not only did they themselves know before the end of the study who was receiving EDTA and who placebo, they had also revealed this information to many of the test subjects. Before the study was over more than 64 percent of the subjects were aware of which treatment they had received. This was highly questionable from an ethical and scientific standpoint.

One important aspect of the Danish study is the startling fact that the patients who were given EDTA were much sicker than the patients treated with a placebo. Therefore, the improvements the EDTA group made were harder earned and more significant. The researchers (who had candidly admitted that they undertook the study to convince the Danish government *not to pay* for chelation) either never noticed that aspect or felt reluctant to reveal it. The evidence is seen in the pre-treatment MWDs: 119 meters for the EDTA group and 157 meters for the placebo group.

Still more significant were the standard deviations. The plus or minus 38 meters SD for EDTA patients versus the plus or minus 266 meters SD for the placebo group represents an enormous variation in walking capacity that is heavily biased in favor of the placebo group. Those standard deviations show that some placebo patients must have walked half a mile before stopping. The placebo group's claudication was there-fore markedly less severe, and the EDTA group was much more severely diseased. The design of the study was obviously biased against EDTA chelation from the outset.

Yet, when the six-month study was completed, the mean MWD in the EDTA group increased by 51.3 percent, from 119 to 180 meters, while the mean MWD in the placebo group increased only 23.6 percent, from 157 to 194 meters. The chela-tion group's improvement was therefore more than twice as great as the placebo group's, even though the chelation group was significantly sicker at the outset. This is a positive study, supporting the usefulness of EDTA chelation. The authors' published negative conclusions are not supported by the data.

The New Zealand Study

Another study—also conducted by vascular surgeons—was done at the Otago Medical School in Dunedin, New Zealand two years after the Danish study. The subjects of this study were also suffering from intermittent claudication. The subjects were divided into two groups, the EDTA group and the control group. The study extended to three months after 20 infusions of either EDTA or a placebo were given. The authors concluded that EDTA chelation had been ineffective. Once again, that conclusion was unsupported by their data.[86,87]

Absolute walking distance in the EDTA group increased by 25.9 percent; while in the placebo group, it increased by 14.8 percent. The difference was not considered statistically significant. The study, however, had only 17 subjects in the placebo group. One of the placebo patients was what the statisticians call an "outlier," one whose results differ strikingly from everyone else in the group. This patient's walking distance increased by almost 500 meters. All of the statistical gain in the placebo group was due to this one individual's progress. Without him, the placebo group's distance actually *decreased*.

This illustrates the perils of a small study. The 25 percent gain in the EDTA group compared to no gain in the placebo group would have been very significant statistically.

In addition, the New Zealand researchers conceded that improvement in artery pulsatility (pulse intensity) in the EDTA group's worse leg improved enough to reach statistical significance ($p<0.001$).

A 25.9 percent improvement in walking is by no means minor and would attract notice if the agent had been a patentable drug. Even that level of improvement is not representative of the much greater improvements claudication patients normally experience after chelation. The below-expected improvement seen in this study can be explained by smoking. Eighty-six percent of the chelated subjects were smokers. Although they were advised to quit smoking when the study began, how many of them actually complied is not known.

The Heidelberg Study

Another study that was carried out with an erroneous negative conclusion is the "Heidelberg Trial," funded by the German pharmaceutical company Thiemann, AG in the early 1980s. A group of patients with intermittent claudication were given 20 infusions of EDTA and compared with a so-called "placebo" group that was actually given bencyclan, a pharmacologically active vasodilating and antiplatelet agent owned by Thiemann.

From a practical commercial standpoint, Thiemann's action was bizarre. If EDTA did well in the trial, Thiemann's well-established drug could only suffer. Nonetheless, the trial went forward and was reported in 1985 at the 7th International Congress on Arteriosclerosis in Melbourne, Australia.[87] Immediately following 20 infusions of EDTA the trial subjects' pain-free walking distance increased by 70 percent. By contrast, patients receiving bencyclan increased their pain-free walking distance by 76 percent. The difference between these two results was not statistically significant, but another result was. Twelve weeks after the series of infusions was completed, the EDTA patients' average pain-free walking distance had continued to increase, going up to 182 percent. No further improvement had occurred in the patients receiving bencyclan. Those percentages were made public, although never published.[87]

An informal report from Thiemann mentioned only the 70 and 76 percent figures. Press releases stated that chelation was no better than a placebo, but failed to mention that the "placebo" was a drug that had been proven effective in the treatment of intermittent claudication. Thiemann never released the actual data on which the Heidelberg Trial based its conclusions, but some German scientists who had access to it, and who were disturbed at the deception they were witnessing, chose to reveal the complete raw data to members of the American scientific community.

The complete data showed that four patients in the EDTA group experienced more than a 1,000-meter increase in their pain-free walking distance following treatment. That highly significant data from those four patients mysteriously disappeared before the final results were made public. Thiemann had a legal right under terms of their contract to edit the final results and to interpret the data in any way that suited them. Another analysis of the data, with the four deleted patients included, showed an average increase in walking distance of 400 percent in the EDTA treated group—five times the 76 percent increase of the group receiving bencyclan.[89]

The Kitchell-Meltzer Reappraisal

A dark moment for chelation research occurred in 1963, when Drs. J.R. Kitchell and L.E. Meltzer coauthored an article reassessing their support for EDTA chelation.

Although it was hardly in widespread use in 1963, chelation had not been controversial. Beginning in 1953, Dr. Norman Clarke and his associates at Providence Hospital in Detroit began using EDTA chelation to treat coronary artery disease. In 1956 they treated 20 patients suffering from angina pectoris. Nineteen of the 20 patients who received EDTA had had a "remarkable improvement" in symptoms.[1]

Soon other physicians became interested, among them Kitchell and Meltzer, at Presbyterian Hospital in Philadelphia. From 1959 to 1963, Kitchell and Meltzer reported good results treating cardiovascular diseases with EDTA. Their early reports were all very positive.[7,10,14]

In April of 1963, shortly after their last favorable report, Kitchell and Meltzer published a "reappraisal" article in the *American Journal of Cardiology* that questioned chelation's value.[75]

In that reappraisal, they reported on 10 of the original patients they had treated for cardiovascular disease, plus another 28 patients that were treated subsequently. Patients in this study were all severely ill. The authors state, ". . . we selected ten patients referred to us because of severe angina. The patients had previously been treated with most of the accepted methods, and their inclusion in this study resulted from wholly unsuccessful courses. Each of the patients was considered disabled at the start of therapy." This was therefore a high-risk group of very sick patients, who had not improved with any other form of therapy.

Seventy-one percent of patients treated had subjective improvement of symptoms, 64 percent had objective improvement of measured exercise tolerance three months after receiving 20 chelation treatments, and 46 percent showed improved electrocardiographic patterns. Kitchell and Meltzer concluded that chelation was not effective because some patients eventually regressed long after treatment. However, considering the poor health of the patients, some eventual worsening would be expected with any treatment. Eighteen months following therapy, 46 percent of the patients remained improved. The results were very favorable, even though the authors' conclusions were not.

Kitchell and Meltzer's reappraisal article was largely responsible for termination of hospital-based, academic research into chelation as a treatment for cardiovascular disease. Rather than analyzing the data for themselves, many physicians simply accepted the flawed conclusion at face value. We will probably never know what prompted those early researchers to change their position so abruptly. We can only speculate that it was an unrealistic expectation that the emergence of bypass surgery would be a final solution.

Cell-Senescence Model of Aging

The cell-senescence model (sometimes called the telomere theory) of aging is now hypothesized to underlie almost all aspects of age-related diseases, including atherosclerotic cardiovascular disease.[90-92] As cells continuously die and are replaced with daughter cells, accuracy of gene expression progressively deteriorates. Replacement cells, produced by cell division and DNA replication, grow increasingly weaker. With each replication, accuracy is lost and subsequent generations of cells reflect that deterioration.

Telomeres on chromosomes shorten with each successive cell division, eventually becoming spent. After 50 divisions, cells reach the so-called Hayflick limit; telomeres become fully depleted and those depleted cells lose their ability to divide. Without telomere replacement, cell death results.

There is more to the story than that. Telomere shortening also correlates with cell senescence, leading to gradual but progressive deterioration throughout multiple generations of daughter cells. When cell deterioration is sufficient to cause impairment in organ function, age-related disease results. It was once thought that only the absolute limit of cell division was important, when telomere length was totally depleted. We now know that as cells divide and telomeres shorten, cumulative inaccuracies in DNA replication cause progressive deterioration of gene expression. This is termed cell senescence. Cells divide and heal more slowly with each successive division.

Two types of cells that do not senesce are germ cells and cancer cells, both of which contain telomerase, an enzyme that restores telomeres. They are thus called immortal cells. When other types of cells, such as fibroblasts, are genetically modified in culture to contain telomerase, they also become immortal and do not senesce. Cell senescence is not only reversed, but aging ceases in cells that produce telomerase, even after 400 or more subsequent divisions.[93]

Endothelial cells in blood vessel walls, lacking telomerase, deteriorate with each cell division.[94] With progressive telomere shortening, cell division and replacement slows and the healing process is retarded. Replacement cells also become increasingly weaker and defective with each division. Cells that are subjected to frequent trauma and injury divide more frequently and therefore age more rapidly. Endothelial cells become disrupted at points of stress, and divide most frequently at precisely those points where atherosclerosis prevails. When endothelial cells are injured and replaced by adjacent cells, daughter cells are produced to fill in the gap. Damage to endothelia occurs with sheer stress at points of bifurcation, from hypertension, infection, toxins, tobacco byproducts, hyperglycemia, oxygen radicals, oxidized LDL cholesterol, and autoimmune processes. The cell senescence model therefore accommodates and provides a broadened explanation for all known risk factors of atherosclerosis.

Senescent endothelial cells divide at a progressively slower rate and are progressively less effective at closing breaches in arterial walls. Prolonged exposure of denuded subendothelial tissues triggers a cascade of events that encompasses current theories of causation: monocytes and platelets adhere to damaged areas; monocytes transform into macrophages; a variety of trophic factors and mitogenic factors, including cytokines and platelet-derived growth factor, are released locally; smooth muscle cells proliferate; oxidized lipids accumulate in macrophages; and eventually this enlarging plaque calcifies or ulcerates. An important underlying cause of this chain of events is postulated to be the progressive telomere shortening in endothelial cells at points where they are most often called upon to divide. This occurs at arterial sites most subject to damage and therefore to plaque formation.

Cell types that divide frequently throughout life are precisely those cells that show age-associated decline. Children with one type of progeria are born with short telomeres. These cells quickly senesce and reach the Hayflick limit of cell division. Skin cells and cells in hair follicles are frequently replaced throughout life. Resulting deterioration is plainly visible to the naked eye, to the extent that a close estimate of chronological age can be made at a glance. Those cell types that divide frequently throughout life and therefore age at predictable rates include chondrocytes, fibroblasts, keratinocytes, microglia, hepatocytes, and lymphocytes. Resulting DNA mutations and reduced immune function can be initiating events in cancer.

Rare or absent division of neuronal cells seems to contradict this theory, since brain function so often declines with age. Astrocytes in the brain, however, continue to divide throughout life and are prominent in the early inflammatory stages of Alzheimer's dementia. Neurons make up only 10 percent of brain cells.

The cell senescence model leads to the conclusion that cell division and telomere shortening are central to cell senescence, and thus to age-related diseases.

To explain actions of EDTA using the cell senescence model, it is only necessary to consider that the lengthy biochemical pathway required for cell division and replication depends on a very large number of metallo-enzymes. DNA-dependent RNA polymerase, an enzyme involved at an early stage of cell replication, is zinc-dependent. Other enzymes needed for replication require the entire spectrum of essential nutritional minerals and trace elements. EDTA has its only known effect on those metallic ions—binding, redistributing, and removing them.

Recent data show that essential trace elements accumulate in diseased tissues. Those elements all have the potential to be toxic in excess. Ischemic myocardium contains a spectrum of essential, nutritional trace elements at quite high levels compared with myocardial cells of healthy, young control subjects: cobalt increases 500 percent; chromium increases 520 percent; iron increases 400 percent; and zinc increases 280 percent.[95] Metallic trace elements have a narrow margin between normal and toxic levels. Such three- to four-fold intracellular elevations could easily poison metabolism. Toxic metals also increase in ischemic myocardium, but less than the essential elements. It is quite possible that by restoring a normal distribution of essential metallic elements within the body, EDTA chelation therapy produces its benefits. This action may be more important than the renal excretion of toxic metals. We just do not know at his time.

Apoptosis (Programmed Cell Death)

When intracellular stresses reach a critical threshold, permeability transition pores (PTP) open in mitochondrial membranes and holocytochrome-C is released. The combination of holocytochrome-C with d-ATP then triggers cellular damage and death. That process is termed apoptosis. Cells become increasingly susceptible to apoptosis with each successive DNA replication and cell division, adding support to the cell-senescence model introduced above. When free oxygen radicals reach high enough levels, apoptosis occurs. Formation of a specific protein, called BAX protein, also opens the PTP, and triggers apoptosis.[96,97] Synthesis of BAX protein requires enzymes that contain such trace elements as cofactors. Every step in the process of apoptosis involves metal-dependent enzymes, and it is on those metals that EDTA has its only known action.

Free Radical Causes of Degenerative Disease

The free radical concept helps to explain contradictory epidemiologic and clinical observations and provides an additional scientific rationale for treatment and prevention of major causes of long-term disability and death such as atherosclerosis, dementia, cancer, arthritis, and other age-related diseases.[98-107] Detection and direct measurement of free radicals has only recently been possible.[108-110] The field of free radical biochemistry is as revolutionary and profound in its implications for medicine as the germ theory was for the science of microbiology. It has created a new paradigm for viewing the disease process. Recent discoveries in the field of free radical pathology, in combination with the cell senescent model and apoptosis, provide a coherent and elegant scientific explanation for many of the reported benefits following EDTA chelation therapy.

Properly administered intravenous EDTA, together with a program of applied clinical nutrition and modification of health-destroying habits, acts synergistically to prevent free radical damage.

What Are Free Radicals?

A free radical is a molecular fragment with an unpaired electron in its outer orbital ring, causing it to be highly oxidative, unstable, and to react instantaneously with other substances in its vicinity.[111,112] The half-life of biologically active free radicals is measured in microseconds.[100] Within a few millionths of a second, free radicals have the potential to react with and damage nearby molecules and cell membranes. Such reactions can then produce an explosive cascade of free radicals in a multiplying effect—a literal chain-reaction of damage.[98,101,102,113,114]

Free radicals react aggressively with other molecules to create aberrant compounds. Harmful effects of high-energy ionizing radiation (ultraviolet light, x-rays, gamma rays, nuclear radiation, and cosmic radiation) are similarly caused by the free radicals produced in living tissues when photons of radiation knock electrons out of orbiting pairs.[105,115-119] Free radicals in cell membranes produce damaging lipid peroxides, oxyarachidonate, and oxycholesterol products.[98,119-122] Oxidized cholesterol is toxic and contributes to atherosclerosis. Lipid peroxides can lead to chain reactions, accelerating a cascade of damaging free radical reactions. Protection against free radical damage is achieved from dietary, supplemental, and endogenous antioxidants.[98,99,101-103,106,121,123,124]

Ongoing free radical reactions in normal cellular metabolism occur continuously in all cells of the body and are necessary for health.[98-103,105-107,125-128] Mitochondrial oxidative phosphorylation produces free radicals during the ongoing production and storage of energy (ATP) from oxygen. These normal and essential free radical reactions are contained and damage is prevented if adequate antioxidant protection is available. The highly reactive free radicals continuously produced within healthy human cells include hydroxyl radicals, superoxide radicals, and excited or singlet state oxygen radicals. They are commonly referred to collectively as "free oxygen radicals," or more simply as "free radicals."[99-107] When free radicals react in the body they in turn produce other highly reactive molecules, including hydrogen peroxide, lipid peroxide, and other peroxides. Peroxides are also metastable, highly reactive, corrosive molecules and also react rapidly, producing additional organic free radicals in surrounding tissues.[111,112]

To prevent uncontrolled propagation of free radicals, cells normally contain a dozen or more antioxidant control systems that regulate the many necessary and desirable free radicals present.[98-106,109,110,116,121,123,129-134] Those control mechanisms include endogenous enzymes, such as catalase, superoxide dismutase, and glutathione peroxidase. Free radical regulation also depends on nutritional antioxidants such as vitamins C and E, beta-carotene, coenzyme Q-10, and the trace element selenium. In fact, almost all vitamins, including the B vitamins, play a role in antioxidant protection. When functioning properly, antioxidant systems suppress excessive free radical production, allowing oxidative energy metabolism to proceed without cellular or molecular damage. When those control systems are weakened, free radicals multiply out of control, much like a nuclear chain reaction, disrupting cell membranes, damaging enzymes, interfering with both active and passive transport across cell membranes, and causing mutagenic damage to nuclear DNA. This is one cause of cancer.[102,109,113,114,120,121,135,136]

Concentration of the free radical control enzyme, superoxide dismutase (SOD), in mammals is directly proportional to life span. Humans have the highest concentrations of SOD. SOD is the fifth most prevalent protein in the human body.[101,102] Elephants, parrots, and other long-lived animals also have high levels. Thus, life expectancy seems to be highly dependent on effective free radical regulation.

Nonenzymatic free radical scavengers are stoichiometrically consumed on a one-to-one ratio when neutralizing free radicals. These include beta-carotene (provitamin A), vitamin E, vitamin C, glutathione, cysteine, methionine, tyrosine, cholesterol, some corticosteroids, and selenium. Once neutralized, other vitamins and enzymes are necessary to restore antioxidant activity, but they must all be present in adequate amounts.

Enzymes involved in free radical protection are proteins, but also require nutritional metallic trace elements or B-vitamins as co-enzymes. For example, copper, zinc, and manganese are all essential for superoxide dismutase activity; selenium is essential for glutathione peroxidase; and iron is necessary for catalase and some forms of peroxidase. Tens of thousands of different enzymes in the body depend on vitamins, trace elements, and minerals to function. Optimum dietary intake of those nutrients is therefore necessary for protection against free-radical mediated, age-related diseases. Recent epidemiological studies show that it is difficult or impossible to receive optimal amounts from food alone, without supplementation.

Identifying Free Radicals

Free radicals exist in very low steady-state concentrations. They rarely reach levels high enough for direct analysis.[98,102]

Sophisticated instruments that allow us to recognize the importance and extent of free radical damage in tissues have only recently become available. Electron paramagnetic resonance spectroscopy (EPR) is one type of technology now used.[109,110] Free radicals can be estimated most easily, and perhaps even more accurately, by analyzing end-products of free radical reactions, using gas chromatography, mass spectroscopy, and high-performance liquid chromatography. Cross-linkages between molecules, damaged collagen, lipid peroxides, oxyarachidonate, oxidized cholesterol, lipofuscin, ceroid, and increased pigment are all caused by free radical reactions. Those substances can be easily measured.[98,101,102,110,137,138]

By sifting through the molecular wreckage left in the wake of evanescent free oxygen radicals, it thus becomes possible to indirectly estimate the type and extent of ongoing free radical reactions. For example, free radical damage in the brain and central nervous system (CNS) can be assessed by the rate of cholesterol depletion. Cholesterol is not otherwise metabolized in the nervous system. The only way for cholesterol to decrease in the CNS is through oxidation caused by free radicals. Cholesterol acts as an antioxidant and is consumed in the process.[101,102,139,140]

Cholesterol Metabolism

As a free radical scavenger, cholesterol is liberally disbursed throughout cell walls and lipid membranes in the body. Contrary to the popular notion that cholesterol is harmful, it actually protects cell membranes—if it has not previously become oxidized in the process of neutralizing a potentially harmful free radical.[110,139] Unoxidized cholesterol is one of the body's important antioxidant defenses. Some of the cholesterol-derived steroid hormones, including glucocorticosteroids, dehydroepiandrosterone (DHEA), pregnenolone, testosterone, progesterone, and estrogen, can all function as free radical scavengers.[110] Those substances decline steadily with age, inversely related to the increase of age-associated diseases.

Cholesterol is a precursor to vitamin D. Vitamin D is normally produced in the skin by exposure of cholesterol to ultraviolet radiation from sunlight. Without cholesterol, vitamin D deficiency can occur. Ultraviolet light is a form of ionizing radiation that can also produce free radicals in the skin manifested as sunburn and skin cancer. Unoxidized cholesterol and other antioxidants act to protect the skin.

Total body cholesterol (approximated by measuring blood levels of cholesterol) is derived primarily from cholesterol synthesis within the liver, not from dietary intake.[102] Blood cholesterol levels increase with free radical stress. Elevation of blood cholesterol may be an indicator of exposure to excessive free radicals and increased risk of atherosclerosis and apoptosis. Also, as cholesterol becomes oxidized, in the form of low-density lipoprotein (LDL) cholesterol, LDL receptor sites in the liver and elsewhere are altered, causing increased hepatic synthesis of endogenous cholesterol. It is possible that an increase in cholesterol is a desirable physiologic response to free radical stress. Cholesterol is synthesized in the body as needed, and the need is greater to protect those at risk. In Western cultures, where atherosclerosis, cancer, and other free radical mediated diseases are epidemic, blood cholesterol levels commonly increase with age.

After encountering and neutralizing a free radical, a molecule of cholesterol is oxidized as LDL cholesterol. In oxidized LDL form, cholesterol is toxic to blood vessel walls. If antioxidant protection is diminished, or if free radical production exceeds the threshold of tolerance, oxidized LDL cholesterol increases and contributes to atherosclerosis.

A recent multi-country study in Europe, funded by the World Health Organization, showed low blood levels of vitamin E are statistically 100 times more significant as a predictor of coronary heart disease than are high blood levels of cholesterol.[141] In another report, all published autopsy studies that attempted to corrclate the extent of atheromatous arterial plaque with levels of blood cholesterol were reviewed. Surgical specimens removed at the time of bypass surgery were also analyzed. After eliminating data from those few individuals with a hereditary form of extremely high cholesterol (above 400 mg/dL) no significant correlation was found between blood cholesterol levels and the severity of atherosclerosis.[142] The author concluded that prior studies had been skewed to a contrary conclusion by failure to eliminate those occasional individuals with extremely high cholesterol caused by a lethal genetic mutation.

Less than one half of one percent of the population has a hereditary trait for dangerously high cholesterol. People with that genetic disease have cholesterol blood levels above 400 mg/dL. They commonly suffer premature death from atherosclerosis despite aggressive pharmacological therapy to lower blood cholesterol. The statements in this section do not apply to people in that category.

Free radicals oxidize cholesterol into a variety of breakdown products.[98,100,101,102,110,143,144] Oxidized cholesterol is bound selectively to low-density lipoproteins, referred to as LDL cholesterol, while unoxidized (antioxidant) cholesterol is predominately bound to high-density lipoproteins, HDL cholesterol.[101,102] Oxidized cholesterol bound to small, dense lipoprotein molecules are especially toxic to cells.

Laboratory research at the Cleveland Clinic demonstrated that both EDTA and the antioxidant glutathione prevent LDL cholesterol from becoming toxic.[143]

Oxidized forms of cholesterol possess varying toxicities.[98,102,143-146] Some of those substances have vitamin D activity, which can cause localized vitamin D toxicity in tissues and macrophages.[144] Abnormal calcium deposits in tissues and blood vessel walls may to some extent be in response to localized vitamin D increase to toxic levels. Free radicals also cause tissue calcification by damaging the integrity of cell membranes, causing leaks in cell walls, and by damaging enzymatic cell-wall transport pumps. If the calcium pump is weakened, or if cell wall integrity is damaged, the calcium pump becomes unable to remove calcium as it leaks in. Intracellular calcium accumulates, causing malfunction and eventually cell death. X-rays of older people commonly show dense calcium deposits in soft tissues that do not normally have that bony appearance. A similar weakening of the sodium pump in cell walls allows an increase of intracellular sodium, leading to swelling of the cell and eventual cell lysis.

Dietary restrictions of cholesterol and prescription drugs to reduce blood cholesterol have, in some ways, been counterproductive because the antioxidant role of cholesterol has not been widely recognized. Natural, unoxidized cholesterol is widely dispersed in cell membranes as a protective factor against atherosclerosis, cancer, and other free radical-induced diseases. In this form, cholesterol is not the harmful substance we have been told. Cholesterol is a fat, and dietary cholesterol is consumed in fatty foods. Restriction of dietary cholesterol necessarily results in simultaneous reduction of total dietary fats. Research studies that allegedly show benefit from low-cholesterol diets may have only reflected benefit from reduction in excessive fat and the accompanying lipid peroxides and oxidized cholesterol, as often occur in processed foods.

It is not widely known that cholesterol-lowering drugs also have antioxidant activity, and antiplatelet activity.[147] Those drugs also produce significant toxicity and cost much more than antioxidant nutritional supplements.

Free radicals cause damage by oxidation. Fats, especially unsaturated fats, are especially susceptible to oxidation, in the same way that cooking fats and oils can easily catch fire on the stove. They ignite (oxidize) easily and burn vigorously. All cells and intracellular organelles are enveloped in easily oxidized layers of unsaturated fat. Damaging oxidation of fatty cellular and intracellular membranes by free radicals can be prevented in three ways:

1) By "fire-proofing" lipid membranes with nutritional antioxidants;

2) By depriving the fire of fuel by partially restricting dietary fats; and

3) By removing metallic catalysts of free radical proliferation with EDTA chelation therapy (as described in detail below).

These three strategies used together act synergistically to reverse and slow diseases of aging.

A statistical correlation has been reported between low blood cholesterol and increased risk of cancer.[148] Cancer is caused in part by free radical damage to nuclear material and chromosomes. Free radicals act as both primary initiators and subsequent promoters of malignant change. If adequate unoxidized cholesterol is not present to provide antioxidant activity for nuclear membranes and DNA, an important defense against mutation and cancer is lost. High-fat diets, rich in lipid peroxides, are known to increase the risk of cancer.[98,101,102] However, if antioxidant defenses are reinforced, dietary fats can be protected against damaging oxidation and rendered useful for energy.

Homocysteine Metabolism

Homocysteine can contribute to atherosclerosis in several ways. The higher the homocysteine level, the greater the risk.

Free radical production and oxidative stress occur during homocysteine metabolism.[149-153] Homocysteine is a metabolite of methionine, and is oxidized by free radicals to become homocysteic acid, a potent stimulator of cell growth and multiplication.[154] In this way, oxidative breakdown of homocysteine can induce proliferation of smooth muscle cells in arterial walls and promote growth of atherosclerotic plaques.[155] Metabolic pathways of homocysteine that cause damage to blood vessel walls involve hydrogen peroxide, superoxide radical, and inhibition of glutathione peroxidase.[156] Homocysteine also increases the tendency for blood clotting.[157] In addition, homocysteine can increase the oxidation of cholesterol, which then becomes bound to small dense LDL particles and is taken up by macrophages to become foam cells in plaque.[158]

Damage to blood vessel walls from elevated homocysteine results in accelerated plaque growth, followed by cholesterol and lipid deposition.[159] Clinical and epidemiological research has shown that atherosclerosis throughout the body correlates with blood levels of homocysteine. Elevated homocysteine is a powerful and independent risk factor, as strong or stronger than other well-documented risk factors.[150-152]

Fortunately, there is an easy solution—vitamin supplementation. Homocysteine is metabolized by enzymes that require vitamins B-6, B-12, and folate as cofactors.[160-161] Intake of those vitamins correlates inversely with homocysteine and atherosclerosis.[162,169] With daily supplementation of B-complex vitamins containing 800 mcg per day of folate, five mg or more of vitamin B-6, and 50 mcg or more of vitamin B-12, even those people with a familial tendency for high homocysteine can prevent the related problems.[162-166]

Essential Free Radical Reactions

Life cannot exist without a balance of carefully regulated free radical reactions. Because of that, life in the presence of oxygen requires antioxidant protection. Cellular respiration requires transfer of electrons across mitochondrial membranes within cells. For every electron that crosses the mitochondrial membrane, a superoxide radical is produced. Antioxidants prevent damage to vital intracellular structures during this process. Humans cannot utilize food for fuel without continuous cellular oxidation-reduction reactions, which produce free radicals as a byproduct. Oxygen is breathed in through the lungs and transported in the blood to every cell in the body, where oxidation reactions produce life-supporting energy. Red blood cells even produce free radicals during the binding and release of oxygen and carbon dioxide by hemoglobin.

Superoxide radicals are released during oxidative phosphorylation of ATP. Cellular protection from the superoxide radicals is provided by mitochondrial superoxide dismutase (SOD), a manganese-containing enzyme. The average American diet contains suboptimal amounts of manganese.[170] SOD in the cytoplasm of cells requires both zinc and copper for activity. Those trace elements are also marginal to deficient in the average American diet.[170] If the integrity of cell membranes is not protected by adequate SOD, then the activity of other vital enzymes contained within cell membranes is compromised.[171]

The metabolism of many chemicals, including most prescription drugs, artificial colorings and flavorings, petrochemicals, and inhaled fumes, takes place in the endoplasmic reticulum of liver cells and other organs. That detoxification process releases hydroxyl free radicals and peroxides.[98,101,102,125-127] Glutathione peroxidase, vitamin C, and other antioxidants must be present in adequate supply to prevent chain reactions of damaging free radicals. In that way, drugs and other chemical exposure cause increased production of free radicals, which may then exceed the threshold of antioxidant protection.[98,101,102] The resulting excess of free radicals may then multiply further, in a chain reaction, magnifying the damage by a million times or more.[98,102]

Synthesis of prostaglandins and leukotrienes from unsaturated fatty acids also results in the release of free radicals.[102,106,128] Lacking sufficient antioxidant protection, prostaglandin production may become imbalanced. For example, production of thromboxane can increase, and prostacyclin can decrease in the presence of lipid peroxides.[98,101,102] Thromboxane is associated with atherosclerosis while prostacyclin acts to prevent arterial plaque.

Leukocytes and macrophages normally produce free radicals. Disease-causing organisms are ingested and destroyed by free radicals during phagocytosis. The leukocytes use free radicals much like "bullets" against an invading army.[172,173] Antioxidants localize and limit the damage caused by the free radicals. If antioxidant protection is inadequate, free radicals migrate into adjacent tissues and produce inflammation, manifested by redness, heat, pain, and swelling.

Without antioxidant enzymes, we would die very quickly. The aging process is accelerated by the age-related decrease in antioxidant protection.[98,101,102] An extreme example of accelerated aging is the disease known as progeria, caused by hereditary absence of free-radical protective enzymes. Within ten to fifteen years after birth, a victim of that genetic mutation will experience every aspect of the aging process, including wrinkled, dried, and sagging skin, baldness, bent and frail body, arthritis, and advanced cardiovascular disease. Administration of an antioxidant has successfully slowed one form of progeria. Heredity determines each individual's unique resistance to free radical mediated disease; therefore, there is a wide variation in tolerance to the dietary and lifestyle stresses that increase free radical production.

Oxygen Toxicity

The process of free radical pathology is often referred to as "oxidative stress." Ground state or unexcited atmospheric oxygen has the unique property of being both a free radical generator and a free radical scavenger.[98,102,174,175] Although a liter of normal atmospheric air on a sunny day contains over one billion hydroxyl free radicals,[122] oxygen at normal physiologic concentrations in living tissues neutralizes more free radicals than it produces.[176] When oxygen concentration falls below normal levels, as occurs with diseased arteries and ischemia, oxygen becomes a net contributor to free radical production.[101,102,177]

Oxygen in high concentrations for prolonged periods of time can cause toxicity and even death, primarily by free radical damage to the lungs and brain. Under proper conditions, however, intermittent high-pressure oxygen administered for short periods in a hyperbaric chamber can stimulate an adaptive increase of intracellular superoxide dismutase, an enzymatic antioxidant.[175,178] Too much or too little oxygen can be equally harmful. High levels of oxygen should therefore only be given in short pulsitile exposures, to stimulate an adaptive increase in antioxidant defenses without causing harm.

Protection Against Oxygen Free Radicals

Normal oxygenation of tissues strengthens defense against free radicals. Aerobic exercise stimulates blood flow, improves oxygen utilization, and increases oxygenation of distal capillary beds. Increased tissue oxygenation during exercise thus acts to protect against free radicals and reduces free radical related disease.

Trace elements are essential components of antioxidant metallo-enzymes. Each molecule of mitochondrial SOD contains three atoms of manganese. Each molecule of cytoplasmic SOD contains two atoms of zinc and one atom of copper.

Each molecule of glutathione peroxidase contains four atoms of selenium. Catalase and peroxidase contain iron. Elemental selenium is an antioxidant, independent of its function as an enzyme co-factor.[102]

The human body lacks an intrinsic defense against one destructive free radical precursor—excited state singlet oxygen. When superoxide radicals exceed a concentration that can be neutralized by SOD, they spontaneously convert to singlet oxygen. Polynuclear aromatic hydrocarbons and aldehydes found in tobacco tar and tobacco combustion products produce free radicals in the body, including singlet oxygen. The most important protection against singlet oxygen is dietary intake of beta-carotene; a precursor form of vitamin A.[179-183] Epidemiological evidence correlates increased dietary beta carotene with reduced incidence of cancer. Fully active vitamin A lacks this protective activity.[101,102]

Beta-carotene is inactivated by free radicals and must be re-activated by other antioxidants, including vitamin C. Vitamin C is inactivated in the process. Cigarette smoke depletes vitamin C. If smokers are given beta-carotene, the oxidized beta-carotene further depletes vitamin C and other antioxidants. It is therefore important to supplement with a full spectrum of antioxidants and micronutrients simultaneously. In one study, an increase in lung cancer was reported when beta-carotene alone was given to smokers. If beta-carotene and vitamin C are not supplemented simultaneously, beta-carotene alone could worsen a preexisting deficiency of vitamin C. That would explain the seemingly paradoxical report of increased cancer with beta-carotene supplementation. Many different antioxidants work together in harmony. An antioxidant is inactivated when it encounters a free radical, and must in turn be reactivated by the next antioxidant in a cascade—a multi-step process.[184-187] All antioxidants must be present simultaneously for optimal protection.

Vitamin E (tocopherol), vitamin C (ascorbate), beta-carotene, selenium in glutathione peroxidase, the amino acid cysteine in reduced glutathione, riboflavin, niacin, and a spectrum of other antioxidants are all interrelated in a recycling process that protects against free radicals. If all of those antioxidants are present in optimum amounts, they are continuously restored to their active forms, after becoming inactivated by free radicals.

The process proceeds as follows: vitamin E or beta-carotene neutralizes a free radical by becoming oxidized to tocopherol quinone or oxidized beta-carotene. Tocopherol quinone and oxidized beta-carotene are recycled into antioxidant vitamin E (tocopherol) or reduced beta-carotene by vitamin C, which is in turn oxidized to dehydroascorbate. Interestingly, the ratio of ascorbate to dehydroascorbate diminishes progressively with age and no species of animal survives when that ratio falls below one to one.[102] This can be corrected by supplementation.

Inactive vitamin C (dehydroascorbate) is reduced to active vitamin C by glutathione peroxidase (which must contain selenium). Glutathione peroxidase is returned to its active form by oxidation of reduced glutathione. A vitamin B-2 (riboflavin)-dependent enzyme, glutathione reductase, returns oxidized glutathione to its active form. Glutathione reductase is reactivated by a vitamin B-3 (niacin)-dependent enzyme, NADH, which is oxidized to become NAD. NAD is metabolized in the electron transport system, passed from step to step down the carboxylic acid (Kreb's) cycle; potentially destructive energy that originated as a free radical is redirected to useful metabolism. Subsequent steps in energy metabolism require every nutritional vitamin, mineral, and trace element.

This stairstep cascade of oxidation-reduction pathways demonstrates that each component depends on an adequate supply of all other components in the chain.[124] A chain is only as strong as its weakest link. This interdependence explains the sometimes-equivocal results that are reported from clinical trials utilizing just one antioxidant. For years medical scientists have been conducting research by supplementing only vitamin E, beta-carotene, selenium, or vitamin C. Although results were often positive,[180-182] benefits would have been much greater had a whole spectrum of essential nutrients been supplemented simultaneously.[184-188] If free radical production in the body exceeds the neutralizing capacity of this system, serious damage to cell membranes, protein molecules, and nuclear material (DNA) results.[98,101,102,111,119,121,123]

A comprehensive understanding of free radical defenses provides a rationale for nutritional supplementation with a wide spectrum of multiple vitamins and trace elements, in safe amounts and in proper physiologic ratios. Although large amounts of water-soluble vitamins are rapidly excreted, transient elevations in tissues are achieved shortly following ingestion, which saturate those tissues and provide additional free radical protection and enhanced metabolic function.[102]

An 18-year nutritional study of thousands of people, published by researchers at the University of California, Los Angeles, showed that daily intake of a multiple vitamin-mineral supplement that contained at least 500 mg of vitamin C could extend average life expectancy by up to six years.[189]

Increased Production of Free Radicals

When free radicals in living tissues exceed safe levels, the result is cell destruction, malignant mutation, tumor growth, damage to enzymes, and inflammation, all of which manifest clinically as age-related, chronic degenerative diseases. Each uncontrolled free radical has the potential to multiply by a million-fold in a chain reaction, much like a nuclear explosion.[98,101,102,111,119,121,123]

Dietary fats, especially polyunsaturated fats, are major sources of pathological free radicals. Double bonds on unsaturated fatty acids combine very readily with oxygen and can produce lipid peroxides. This occurs both after consumption and while exposed to the atmospheric oxygen prior to consumption.

Lipid peroxidation begins when fats and oils are exposed to air, and is greatly accelerated when heated. Oxidation of fats and oils is catalyzed and hence accelerated enormously by metallic ions, especially free iron and copper. For example, peanuts crushed to make peanut butter are rich in both iron and copper, which are freed into the unsaturated oil when the peanut is disrupted. Iron and copper are potent catalysts of lipid peroxidation and increase the rate of rancidity of peanut oil by up to a million times. Oxidized fats and oils are commonly called rancid; however, extensive peroxidation can exist in some oils without a detectable rancid odor or taste.[102] Lipid peroxidation occurs during the manufacture of many foods and cooking oils.[102]

The more unsaturated the oil in fatty acids, the more readily peroxidation will occur. The rate of peroxidation is proportional to the square of the number of unsaturated bonds in each fatty acid molecule. Factors that further increase the rate of peroxidation include heat, oxygen, light, and trace amounts of unbound metallic elements.[102,119] Oils prepared in the dark, at low temperatures, in an atmosphere of pure nitrogen, and with added fat-soluble antioxidants such as vitamin E, would be best for nutritional use.[102] Such oils are not commercially feasible. The alternative is to eat oil-containing foods, such as nuts and seeds, in their natural state, without crushing until chewed and swallowed. Dietary supplementation with insurance doses of antioxidants can help to prevent free radical damage in the body, even when oxidized fats and oils are eaten.

Unsaturated vegetable oils often contain trace amounts of iron and copper and are routinely exposed to heat and oxygen when foods are fried. That creates the worst possible combination. Hence, the admonition to limit consumption of fried foods. Oils used in the manufacture of salad dressings, such as mayonnaise, often contain high concentrations of lipid peroxides. The poorest quality oils are commonly used to produce commercial food products because heavy seasoning masks rancidity.[102]

Peroxidation and hydrogenation of vegetable oils during the manufacture of margarine and shortening results in cis- to trans-isomerization. Trans-isomerization alters the three-dimensional configuration of dietary fatty acid molecules from their normal "cis" coils to straightened "trans" configurations. Trans-fatty acids are then incorporated into cell membranes in the place of naturally occurring cis forms, weakening the membrane structure and impairing function of phospholipid-dependent enzymes imbedded in cell membranes.[101,102,218,219] Substrate recognition by enzymes that synthesize cell membranes is not adequate to distinguish between these two forms.[102]

Phospholipids that compose cell membranes are easily damaged by free radicals, as already explained. Dietary intake of peroxidized fats initiates that process. The prostaglandin precursors—arachidonic and linoleic acids—are depleted in the process, as measured by gas chromatography.[101,102] Cell membranes containing trans-fatty acids have impaired fluidity and increased permeability that interfere with transport of sodium, potassium, calcium, magnesium, and other substances. Receptors for insulin and other hormones are disturbed.[98,101,102] Damage from trans-isomerization of fatty acids is cumulative with lipid peroxidation.[98,101,102]

Very little attention is commonly paid to the quality of dietary fats and oils. Emphasis has mistakenly been placed on the ratio of saturated to unsaturated fatty acids, irrespective of lipid peroxidation and trans-isomerization. Contrary to conventional wisdom, unsaturated fats are more toxic than saturated fats.[102] Margarine contains far more peroxides and trans-fatty acids than butter. Moderating consumption of all fats and oils is desirable.[102] It has been shown epidemiologically that margarine consumption is a risk factor for heart disease, contrary to advertisements for preventive benefit.[192]

The quantity and quality of dietary fat are as important or more so than the ratio of unsaturated to saturated fatty acids.[98,101,102] If dietary fats and oils are obtained from fresh, whole, unfractionated, and unprocessed foods, they will be minimally oxidized and will produce healthy cell membranes with normal cis-fatty acid configurations. They will preserve a normal balance of prostaglandins. Although fully saturated fats are not as easily oxidized, all animal fats contain some unsaturated fatty acids and cholesterol, both of which are subject to oxidation. Animal experiments have shown that as little as one percent of dietary cholesterol consumed in oxidized form can contribute to atherosclerosis. Supplemental antioxidants reduce that risk.[144,145,193]

How much dietary fat and oil can one tolerate without risk? Evidence indicates that between 25 to 35 percent of dietary calories as fat can be both safe and nutritious, if attention is paid to the quality and source of the fat, as described above, and if supplemental vitamins, minerals, and trace elements are used.[101,102] The more oxidized the fats, the less well they are tolerated. In the United States, an average of 45 percent of dietary calories are consumed as fat, mostly of poor quality, with no consideration for rancidity or cis- to trans-isomerization. Half the population takes little or nothing in the way of nutritional supplements. Lipid peroxidation occurs much more slowly when foods are frozen.[102]

Free radical damage contributes to senility, dementia, and other nervous system diseases, including Alzheimer's and

Parkinson's syndromes. The brain and spinal cord contain the highest concentration of fats of any organ. Nervous system fats are also very rich in highly unsaturated arachadonic and docosahexanoic acids. Because the rate of lipid peroxidation increases exponentially with the number of unsaturated carbon-carbon double bonds, docosahexanoic and arachadonic acids peroxidize many times more readily than most other lipids. The brain and spinal cord therefore require added antioxidant protection.

To provide that extra protection, vitamin C is concentrated in the brain by metabolically active pumps in the blood-brain barrier. Ascorbate is 100 times more concentrated in the brain and spinal cord than in other organs.[102] Two ascorbate pumps operate in series. The first increases the concentration ten-fold from blood to cerebrospinal fluid. A second pump concentrates vitamin C by another factor of ten between cerebrospinal fluid and the subdural space. The disappearance rate of vitamin C from spinal fluid can be used to indicate the extent of damage and subsequent lipid peroxidation following ischemia or trauma to the central nervous system.[194,195]

Experimental spinal cord injuries in animals have been used to show benefit from treatments based on free radical protection. A minor contusion to the spinal cord results in rapid breakdown of unsaturated fatty acid sheaths surrounding nerve pathways. Following a contusion, capillaries leak blood. Erythrocytes hemolyze, releasing free iron and copper. Those metals are potent catalysts and react with oxygen to massively increase the rate of lipid peroxidation.[102] The chain reaction of oxidative damage and inflammation that occurs has been extinguished experimentally in two ways in animals: 1) The spinal cord has been exposed and irrigated with a potent free radical scavenger, such as dimethyl sulfoxide (DMSO); and 2) iron and copper catalysts have been inactivated by bathing the injured area in a chelating solution containing EDTA.[102,109,140,191,196-198]

Intravenous DMSO in large doses has been reported to prevent paralysis and permanent damage in victims of spinal cord and brain injuries. Results thus far indicate that if treatment is begun within the first thirty minutes, or at most within the first two hours, the outcome is much better than would otherwise have been expected.[199-214]

Chelation therapy with EDTA combined with dietary fat restriction has been reported to alleviate or temporarily reverse the progression of multiple sclerosis (MS).[34,215,216] MS victims experience degeneration of the fatty myelin sheaths that insulate nerve pathways in the brain and spinal cord. Although lipid peroxidation may be only one link in the chain of cause and effect, it is sometimes possible to slow this devastating disease by using treatment principles that reduce free radical pathology.

Tobacco and Alcohol

Habitual tobacco use and excessive alcohol use lead to disease and premature death. Alcohol causes damage and scarring to the liver and cancer in the mouth and digestive organs.[217] Alcohol is metabolized to acetaldehyde, which is a potent free radical precursor. Acetaldehyde is closely related to formaldehyde (embalming fluid), and causes cross-linkages of connective tissue by free radical reactions (like tanning leather).[101,102,218,219] Alcoholic cirrhosis of the liver might therefore be considered quite literally a form of pre-death embalming.

Tobacco smoke contains polynuclear aromatic hydrocarbons, which are potent precursors of free radicals. Those substances can easily overwhelm the body's free radical defenses and speed the onset of cancer, atherosclerosis, and other age-related degenerative diseases.[220] Processed tobacco also contains cadmium, a heavy metal ten times more toxic than lead, which acts as a catalyst of free radical reactions and displaces zinc in metallo-activated enzymes.

Cell Membrane Metabolism

Every cell in the body is enveloped in fat in the form of bipolar, phospholipid membranes. Spanning those membranes are large proteins, enzymes, molecular pumps, and receptors for various hormones and peptides. Cell membranes are metabolically very active and have the characteristics of a viscous fluid. They are constantly changing. Cell membranes have one-way permeability to substances that must be kept out of or inside of cells. The water-soluble, polar ends of phospholipid molecules line up on the inner and outer surfaces of the cell membrane, bathed in aqueous fluids. The fat-soluble, nonpolar tails point toward the center of the membrane, intertwining with the fatty tails of similar molecules extending from the opposite surface, traversing the interior of the cell-wall membrane. The normal, curly cis-configuration of membrane phospholipids allows them to twist around each other and grasp cell-wall proteins, enzymes, and other constituents within the membrane, holding them in proper position. The cis-curvature of healthy lipids is essential to integrity and metabolic activity of cell membranes.[98,101,102]

Unoxidized cholesterol is widely disbursed within cell membranes and acts as an antioxidant.[102] Oxidized dietary cholesterol produced in food processing offers no such protection and is atherogenic.[102,144] When free radicals occur in the vicinity of a cell, unoxidized cholesterol, vitamin C, vitamin E, coenzyme Q-10, and the entire array of antioxidant defenses are needed to prevent damage.

Large enzymatic proteins span the full thickness of cell membranes and act as metabolically active "pumps." They are bathed in plasma on the exterior and extend into the cytoplasm within the cell. One such pump keeps sodium ions out of the cell and potassium within, against a powerful diffusion gradient. Another keeps calcium out and magnesium in. Cellular organelles, including mitochondria, lysosomes, endoplasmic reticuli, Golgi bodies, and nuclear DNA are enveloped in similar bipolar lipid membranes that also contain many energy-dependent transport mechanisms. Mitochondrial membranes are protected by coenzyme Q-10, an antioxidant necessary for safe energy production. Mitochondria are the power plants of cells, and they continually produce free radicals during transport of electrons, in the process of oxidative phosphorylation. Damage to mitochondria causes premature aging. A spectrum of antioxidants is necessary to prevent those free radicals from destroying mitochondria.[98,101,102]

Receptors on cell membranes for neurotransmitters, insulin, hundreds of different oligopeptide regulators, and hormones are also subject to damage by free radicals. The calcium-magnesium and sodium-potassium pumps become weakened with advancing age, allowing harmful levels of calcium and sodium to enter cells. Free radicals damage nuclear membranes, altering nuclear pores and chromosomes, and causing mutations with resulting impairment of protein synthesis and cell replication. Free radical mutations of DNA can cause uncontrolled cell division and cancer.

Free radicals increase the activity of guanylate cyclase, which can stimulate uncontrolled cell multiplication. Lymphoid tissues are very rich in unsaturated fatty acids, and free radical damage can easily cause immunologic abnormalities.[102] The immune system may subsequently attack the body's own tissues in so-called autoimmune diseases or it may weaken and fail to recognize and destroy disease-causing organisms and malignant cells.[98,101,102,106]

Calcium Metabolism

Free radical damage to the calcium-magnesium pump allows excessive calcium to diffuse into the cell. Calcium is 10,000 times more concentrated outside than inside of cells. The calcium pump must constantly work against this gradient. The reverse is true of magnesium. If the pump cannot prevent calcium from leaking into cells, and keep magnesium from leaking out, the cell is poisoned and soon dies.

Calcium activates phospholipase-A2, which cleaves arachadonic acid from membrane phospholipids. Increased levels of arachadonic acid can in turn create an imbalance of prostaglandins and leukotrienes, creating more free radicals in the process.[101,102] Leukotrienes are potent mediators of inflammation and attract leukocytes. Leukocytes, as noted previously, release superoxide free radicals during phagocytosis. Leukocytes, stimulated by leukotrienes, can overpower local antioxidant defenses, leading to inappropriate inflammation and damage to surrounding tissues.[101,102,227] Small capillaries and arterioles dilate, causing edema and leakage of erythrocytes through blood vessel walls. Platelets produce microthrombi. Erythrocytes hemolyze, releasing free copper and iron, which catalyze progressive oxidative damage to adjacent tissues. This results in an inflammatory chain reaction, beyond normal control mechanisms.

Free radical damage causes calcium to leak into smooth muscle cells in arterial walls. Calcium binds to calmodulin, activating myosin kinase, which in turn phosphorylates myosin. Myosin and actin constrict, causing the muscle cells to shorten. By this mechanism, excess calcium within arterial smooth muscle cells causes spasm. The same occurs in cells of the myocardium. When muscle fibers encircling arteries constrict, blood flow is reduced. Calcium channel blockers relieve symptoms by slowing abnormal entry of calcium into cells, but they do not correct the underlying cause of the problem—free radical disruption of cell membranes.[101,102,222]

Myocardial cells are weakened by excessive intracellular calcium, lowering the efficiency of oxygen utilization and placing an extra burden on an already impaired coronary artery system. If a coronary (or other) artery is partially occluded by atherosclerotic plaque, a small amount of spasm superimposed on a preexisting partial blockage can easily cause ischemia. A myocardial infarction can result from spasm alone, even in a coronary artery completely free from plaque.[223] Thromboxane and serotonin are released by platelets in the presence of free radicals.[101,102] Thromboxane and serotonin also cause arterial spasm.

Excessive intracellular calcium can result from a variety of other factors. Ionized plasma calcium, that metabolically active fraction not bound to protein, slowly increases with age. The higher the concentration of ionized calcium outside a cell, the harder the calcium pump must work to prevent excessive calcium from leaking through. Naturally occurring calcium channel antagonists can slow the calcium influx. These include dietary magnesium, manganese, and potassium. Magnesium and manganese intakes are suboptimal in the average American diet.[170] Diets are rarely deficient in potassium, but an excessively high ratio of dietary sodium to potassium is common, allowing excessive sodium to diffuse into cells, weakening metabolism. Losses of potassium and magnesium from diuretic therapy can potentiate this problem.

The efficiency of energy metabolism can be impaired by stress. Stress increases circulating catecholamines and inhibits production of ATPase. The calcium and sodium pumps both require ATPase. Cells lose potassium and magnesium and retain calcium and sodium at a greater rate under stress because of relative inhibition of both magnesium-calcium ATPase and sodium-potassium ATPase.

Catecholamines produce free radicals when they are metabolized.[101,102] If free radicals in the central nervous system exceed defenses, stress-related catecholamines cause free radical damage to neuron receptors. That is a partial explanation for stress-related nervous disorders. Breakdown products of dopamine, a catecholamine neurotransmitter, can also cause free radical damage to neuronal receptor sites in the brain. That is hypothesized to be one factor in Parkinson's syndrome and some types of schizophrenia.[224] Neuronal receptors for norepinephrine can also be damaged by free radicals, leading to depression. Heart disease has also been shown to result from catecholamine-induced free radicals.[225]

In recent animal experiments, primates subjected to stress were found to have an increased incidence of atherosclerosis, even while fed a diet that is otherwise protective. Increased free radical pathology associated with increased catecholamine metabolism provides an explanation.

Dementia of the Alzheimer's type is thought by some authorities to be caused by brain cell destruction from free radicals.[204] Arrest or improvement in that condition has been reported following treatment with deferoxamine, an iron chelating agent.[226] EDTA also removes iron and aluminum very effectively. Free, unbound iron is a potent catalyst of lipid peroxidation. Accumulations of aluminum, combined with lipid and protein breakdown pigments are found in brains affected by both Parkinson's and Alzheimer's, although it is not known whether those are late events and not causative factors—much like the accumulation of calcium and cholesterol in arterial plaque.

Before the significance of free radical pathology was recognized, it was hypothesized that EDTA chelation therapy had its major beneficial effect on calcium metabolism. It now seems more probable that calcium is just another link in the chain of cause and effect created by free radical damage. EDTA can influence calcium metabolism in many ways, but direct action on calcium is not adequate to explain the many benefits that follow chelation.

EDTA lowers ionized plasma calcium during infusion. The body then attempts to maintain a homeostasis by producing parathormone.[227] The intermittent three- to four-hour pulses of increased parathormone caused by EDTA can have a measurable effect on bone metabolism.[228] Frost's concept of bone metabolism, known as the Basic Multicellular Unit (BMU) theory, is accepted by other experts.[229] The BMU theory helps to explain the causes and treatment of osteoporosis and osteopenia.

The BMU is a group of metabolically active cells that control the turnover of approximately 0.1 cubic millimeter of bone tissue. When a BMU is activated, it goes through a cycle consisting of an initial three to four weeks of bone absorption (osteoclastic phase) followed by a two- to three-month period of bone reformation (osteoblastic phase). Net increase or decrease in bone density at the end of the entire three- to four-month cycle depends on the rate and completeness of bone turnover. Hormone regulation of BMUs also includes calcitonin, growth hormone, thyroxin, and adrenal corticosteroids, but parathormone remains the most important controlling factor.[228]

Chronically high levels of parathormone have long been known to cause net bone destruction; but brief pulsatile increases in parathormone, as occur during intravenous EDTA chelation therapy, cause an increase of new bone formation.[52,230]

Anabolic activation of BMUs by pulsatile parathormone secretions provides one possible hypothesis for delayed benefit seen in chelation patients, although the theoretical explanations in this chapter are more likely to be important. It is theorized that calcium deposits might be removed from arteries and other soft tissues for utilization in new bone formation.

In their original studies, Meltzer and Kitchell used EDTA to treat ten men who were severely disabled by heart disease and suffered from intractable angina. After approximately twenty infusions of EDTA, therapy was discontinued because of initially disappointing results. Three months later, nine out of ten patients returned to report marked relief of angina, despite no change in their lifestyles, such as altering smoking or nutritional habits.[7] This three-month delay in achieving full benefits has remained a consistent observation by chelating physicians over the years. The delay in achieving full benefit suggests that EDTA activates some type of long-acting healing mechanism.

As an aside, postmenopausal women who are not supplemented with estrogen experience a large increase in follicular stimulating hormone (FSH). Elevated FSH interferes with new bone formation by BMU cells and is regarded as a contributing factor in postmenopausal osteoporosis.[231]

Before free radical pathology was discovered, it was hypothesized that removal of calcium from atherosclerotic plaque and from pathological cross-linkages could explain most of the benefits seen from chelation. Correction of molecular cross-linking may also be an important benefit from EDTA. Undesirable cross-linkages include disulfide bonds caused by free radical reactions that increase with age, and from intermolecular bridging by divalent cations that increase with age, such as calcium, lead, cadmium, aluminum, and other metals. EDTA can remove those metals. Chelation can also reduce abnormal disulfide cross-linkages. Reduction of cross-linking between molecules acts to restore the elasticity of vascular walls and other tissues that is lost with age (covered in detail in chapter 3).[36,37]

Improvements in blood flow may not be detectable on arteriograms, despite marked clinical improvement. Research shows that arteriograms are limited, and can only measure the diameter of an artery to within plus or minus 20 percent (as described in the introduction). Although counterintuitive, it is true that with perfect laminar blood flow, a mere 19 percent increase in the diameter of an artery doubles flow of blood. In a plaque-filled vessel with turbulent flow, less than 10 percent increase in diameter will result in a doubling of blood flow. This can be proved using Poiseuille's law of hemodynamics, as explained in textbooks of medial physiology.

In an organ with compromised circulation, a 25 percent increase in blood flow could bring significant functional improvement and relief of symptoms. Changes in diameter of such small magnitude cannot be detected on either arteriograms or ultrasound imaging.

EDTA Chelation Therapy

EDTA can greatly reduce the release of free radicals.[102,143,232] It is not possible for free radical reactions to be catalyzed and thus accelerated by metallic ions in the presence of EDTA. Traces of unbound metallic ions are necessary as catalysts of uncontrolled proliferation of free radicals in tissues. EDTA binds those ionic metals, making them chemically inert and rapidly removing them from the body. The amount of metal ions necessary to catalyze lipid peroxidation is so minuscule that the tiny traces remaining in distilled water can initiate and accelerate those reactions.[102,176] Metals incorporated into metallic enzymes are tightly bound and not very accessible to EDTA. Some essential elements are briefly removed, however, and require supplementation for replacement.

To catalyze lipid peroxidation, a metallic ion must easily change electrical valence by one unit. Two essential nutritional elements, iron and copper, are potentially the most potent catalysts of lipid peroxidation, although copper is not prevalent enough in the body to be clinically important in that regard. With age, catalytic iron accumulates adjacent to phospholipid cell membranes, in joint fluid, and in cerebrospinal fluid. It is released into tissues following trauma and ischemia. This unbound form of extracellular iron causes free radical tissue damage, evidenced clinically as inflammation.[110,119,121,233-238]

Toxic Heavy Metals

EDTA has long been the accepted treatment for heavy metal poisoning. Toxic heavy metals can impede metabolism in a variety of ways. Poisonous metals such as lead, mercury, and cadmium react avidly with sulfur-containing amino acids on protein molecules. When lead reacts with sulfur on the cysteine or methionine moiety of an enzyme, enzyme activity is reduced or destroyed. Lead also displaces zinc in zinc-dependent enzymes. Chelation therapy reactivates enzymes by removing those toxic metals. The average concentration of lead in human bones has increased by approximately one thousand fold since the Industrial Revolution.[239] Bone lead is in equilibrium with other vital organs and is released into the circulation with a fever and under stress, increasing toxicity when it can be least tolerated.[240]

Lead destroys the antioxidant properties of glutathione and glutathione peroxidase. Lead reacts vigorously with sulfur-containing glutathione and prevents it from neutralizing free radicals. As previously described, reduced glutathione is an essential antioxidant in the recycling of vitamins E and C, glutathione peroxidase, glutathione reductase, and NADH. Lead therefore poisons free radical protective activity of the entire cascade of antioxidant protection.

Lead reacts with selenium even more avidly than sulfur, inactivating the selenium-containing enzyme, glutathione peroxidase. In addition to performing its role in the antioxidant recycling system, that enzyme protects directly against lipid peroxides. Other toxic heavy metals also inactivate glutathione peroxidase. Testing for levels of those metals, as well as clinical evaluation for adequacy of trace element nutriture, is important in the initial evaluation of a patient prior to chelation therapy (covered in Chapter 38).[170,241-266]

The wide variety of benefits reported following EDTA chelation therapy can be understood using the above concepts. EDTA cannot easily chelate metallic ions when they are tightly bound within metal-containing enzymes and metal-binding proteins. On the other hand, when metals accumulate in unbound form, able to act as catalysts of uncontrolled lipid peroxidation, EDTA can easily bind and remove them. Iron accumulates with age at abnormal locations, where it accelerates free radical damage.[34,233-238] EDTA binds much more tightly to iron and other potential free radical catalysts than it does to calcium. EDTA will only bind calcium if those other ions are not present.[34]

Iron accumulates more slowly in women during the child-bearing years because of monthly menstrual losses. Because of that monthly iron loss, younger women have significant protection against atherosclerosis. That protection is lost at menopause. Body iron stores, best reflected by serum ferritin and transferrin saturation, accumulate in men four times more rapidly than in premenopausal women.[267] The risk of atherosclerosis is also four times greater in men in this same age group. Data from the Framingham study show that two years after a hysterectomy, a woman's risk for cardiovascular disease becomes equal to a man's, with or without hormone replacement. This occurs even if the ovaries are not removed.[268-270] Those observations indicate that slower iron accumulation is primarily responsible for reduced atherosclerosis in premenopausal women, although estrogen also has antioxidant benefit.[271] Periodic donation of blood to the blood bank significantly prolongs life expectancy.[272] The fact that iron is a potent catalyst of lipid peroxidation provides a link between these clinical and epidemiologic findings. EDTA has a very high affinity for unbound iron and rapidly removes it from the body.

The affinity of EDTA to bind various metals at physiologic pH, in order of decreasing stability, is listed below. In the presence of a more tightly bound metal, EDTA releases metals lower in the series and binds to the metal for which it has a greater affinity.[273] Calcium is near the bottom of the list, while iron and toxic metals are near the top.

Chromium 2+

Iron 3+

Mercury 2+

Copper 2+

Lead 2+

Zinc 2+

Cadmium 2+

Cobalt 2+

Aluminum 3+

Iron 2+

Manganese 2+

Calcium 2+

Magnesium 2+

In clinical practice, chromium, mercury, and copper are not removed in any significant amount by EDTA, indicating that they are already more tightly bound in the body than by EDTA.

Magnesium is a calcium antagonist and is relatively deficient in many chelation patients. Magnesium is the metallic ion least likely to be removed by EDTA. In fact, EDTA is usually administered as magnesium-EDTA, which provides an efficient delivery system. It increases magnesium stores and reduces likelihood of pain at the infusion site.

Lasting inhibition of disease-causing free radicals by EDTA offers an explanation for the data from Switzerland, which documented a 90 percent reduction in deaths from cancer in a large group of patients who were chelated and then carefully followed over an eighteen-year period (Chapter 12). Chelation patients were compared with a statistically matched control group. Death rate from cancer was ten times greater in the untreated group, compared to the death rate of patients who had been previously treated with EDTA (P=0.002).[38] A greatly reduced incidence of cardiovascular deaths was also observed. One common denominator of both cancer and atherosclerosis is free radical oxidative damage to molecules.[98,101,102] Calcium-EDTA was administered in that study, which precludes any direct effect on calcium metabolism as an explanation for outcomes. Removal of free radical catalysts seems the likely explanation. Demopoulos first proposed that chelation be used to control free radical pathology.[102,121] He also pointed out that many antioxidants have chelating properties.[98]

EDTA increases the efficiency of mitochondrial oxidative phosphorylation and improves myocardial function, quite independently of any effect on arterial blood flow.[274] Treatment with deferoxamine, an iron chelator, has been shown to improve cardiac function in patients with increased iron stores.[275] In addition, removal of iron with deferoxamine reduces inflammatory responses in animal experiments.[276] Sullivan has suggested that periodic donation of blood be studied as a way to reduce the risk of atherosclerosis in men and postmenopausal women (see Chapter 2).[271-272]

By reducing damage from free radicals, EDTA chelation therapy can support normal healing. The time required for healing of damaged tissues gives us another explanation for the time lapse of several months following chelation before full benefit is achieved. By correcting the underlying cause of the disease process, and allowing time for subsequent healing, treatment with EDTA seems far superior to the mere suppression of symptoms achieved with so many other therapies.

Chelation and Atherosclerosis

If an injury results in bleeding, homeostatic mechanisms quickly stop the flow of blood to prevent hemorrhage and death. This regulation of blood loss is under the control of a complex array of mechanisms, including hormones, prostaglandins, fibrin, and thromboplastin. Prostaglandins are produced and degraded continuously and very rapidly in endothelial cells and platelets. Prostaglandins have a half-life measured in seconds and must be constantly synthesized at a controlled rate and with a proper ratio between their various subtypes to maintain normal blood flow.

The two most important prostaglandins in that regard are prostacyclin and thromboxane. Prostacyclin reduces the adhesiveness of platelets, facilitating free flow of blood cells and plasma, and reducing the tendency for fibrin deposition and thrombus formation. Prostacyclin relaxes encircling muscle fibers in artery walls, reducing spasm. Thromboxane does the opposite. It causes intense spasm in blood vessel walls and stimulates platelets to adhere.[277] In oversimplified terms, thromboxane may be considered as undesirable and prostacyclin as desirable. In actual fact, a proper balance must be maintained to protect against injury and hemorrhage, on the one hand, and to maintain normal circulation, on the other.

Synthesis of prostacyclin is greatly inhibited by lipid peroxides and free radicals, while thromboxane production remains unaffected. If lipid peroxides are present, either from dietary intake of peroxidized fats and oils or from nearby peroxidation of lipid cell membranes, less prostacyclin is produced to balance the effects of thromboxane.[101,102,278]

Ongoing damage to vascular endothelium occurs continuously in response to hemodynamic stresses. Damage may also be caused by disordered immunity or bacteria. With good health, minor vascular injuries are rapidly healed, initiated by a layer of platelets that coat the disrupted surface with a protective blanket.[145] If free radical protection is inadequate and local controls have been taxed, the local increase of free radicals blocks the production of prostacyclin. Without prostacyclin, thromboxane is unopposed and causes the injured area of the arterial wall to attract platelets abnormally. This causes platelets to increasingly stick to each other. Platelets thus aggregate, and the growing layer of platelets traps leukocytes, which in turn produce more free radicals. A network of fibrin and microthrombi is formed, and erythrocytes become trapped. Some of the erythrocytes hemolyze, causing iron to be released. Catalytic iron then produces an explosive increase in free radical oxidation, oxidizing any cholesterol present and damaging phospholipids in cell membranes. Prostacyclin production continues to be inhibited for some distance along the blood vessel.

The resulting high concentrations of free radicals can damage nuclear material in arterial cells, causing mutation and uncontrolled cell replication. Lipid peroxides increase the activity of guanylate cyclase, which speeds mitosis.[98,101,102] Platelets release growth factors. This sequence of events can produce an atheroma, an enlarging tumor consisting of mutated, rapidly multiplying, multipotential cells that have lost their high degree of differentiation and specialized function. Atheroma cells produce substantial amounts of connective tissue, collagen, and elastin. They also act as macrophages, ingesting cellular debris and oxidized cholesterol. The monoclonal theory of atheroma formation first proposed by Benditt[279] most accurately fits these known facts. Cholesterol is oxidized by free radical activity, and some of the cholesterol oxidation products ingested by atheroma cells have vitamin D activity.[144]

We have thus far explained how intracellular calcium increases abnormally because of free radical damage to homeostatic mechanisms. Localized excesses of vitamin D activity are caused by free radical oxidation of cholesterol, analogous to the way sunlight creates vitamin D from cholesterol on the skin. Localized increases in vitamin D activity further accelerate calcium accumulation as plaques grow. Calcium and cholesterol deposition do not occur until late in the process of atheroma formation. In its role as an antioxidant, cholesterol acts to protect against further free radical damage but becomes oxidized in the process. Some cholesterol is synthesized within atheroma cells.[280] Oxidized cholesterol and cholesterol esters thus accumulate within plaque. The plaque gradually expands to exceed its blood supply and ulcerates. When ulceration occurs, the central core of the plaque degenerates into an amorphous fibro-fatty mass containing varying amounts of calcium, cholesterol, connective tissue, and cellular debris. This necrotic core can rupture, releasing embolic showers of plaque debris. Free radicals continue to suppress prostacyclin, causing further aggregation of platelets. Platelets release high concentrations of thromboxane and serotonin, leading to arterial spasm and ischemia.

Symptomatic ischemia usually does not occur until a blood vessel becomes 75 percent occluded. A meal laden with peroxidized fats can cause a sudden free radical insult, triggering an abrupt increase in spasm or even an acute thrombosis, superimposed on a partial occlusion, producing an infarction.

Cell damage of this type may occur in any part of the body. Cells swell and die as membranes become leaky and damaged. Membrane pump mechanisms become uncoupled or disabled. DNA damage results in mutations, atheromata, and cancer.[98,101,102] Lymphoid tissues and other cells of the immune system become damaged.[98,101,102] Tissues become stiffer and lose flexibility as cross-linkages occur in connective tissue, elastin, and protein molecules.

Tissues damaged in this way age more rapidly and organ functions deteriorate. Joints become hypertrophic, inflamed, and deformed with arthritis. Leukotriene production and prostaglandin imbalances cause inflammatory change in joints and other organs. Lysosomes rupture, releasing proteolytic enzymes that devastate cell contents. Lysosomes have been called the cells' digestive organs and, when disrupted, cause cellular autodigestion. Free radicals inactivate selenium, creating inert selenium compounds and causing metabolic selenium deficiency. Cancer patients excrete selenium in amounts up to five times the normal rate, just when they need it the most.[170]

Antibody production and cellular immunity are impaired by free radicals. Cells of the immune system are especially rich in unsaturated fats and are therefore more vulnerable to free radical damage. Oxidized cholesterol and lipid peroxides are potent immunosuppressants.[101,102,106,144] Antigenic substances and malignant cells, which would otherwise be neutralized, can overwhelm a weakened immune system. Intact food molecules that leak across the gut wall undigested are poorly tolerated.[281-283] Adverse reactions to specific foods (so-called "food allergies") then appear. Normal free radical reactions in macrophages during phagocytosis of antigens grow out of control and cause inflammation. Adverse reactions are triggered by a variety of nutritious foods and environmental exposures to which the immune system becomes sensitized. This is an increasingly common cause of symptoms. Avoidance of sensitizing foods and other trigger factors becomes necessary to control symptoms.[284-288]

Antigenic properties and toxins released by *Candida albicans*, a yeast normally present in the body in small numbers, can overwhelm the immune system.[289-295] A struggling immune system may become over-reactive in other areas, attacking healthy tissues, leading to so-called autoimmune syndromes.

Atheroma and Cancer: Both Are Tumors

The development of cancer can take decades from the initiating event to the onset of symptoms. If cancer-promoting factors are removed, free radical damage can be repaired and healing can be aided by a more nutritious diet, antioxidant supplementation, and lifestyle improvements. In the early stages, malignant cells have the ability to transform back into a normal, benign state. For example, smokers who stop the use of tobacco have approximately the same risk of cancer ten years later as those who never smoked.[102]

Atherosclerotic plaque is actually a benign tumor, an atheroma, somewhat analogous to cancer. It does not metastasize but expands locally to occlude the flow of blood. It can regress with time, if causative factors are removed. Free radical pathology is the common denominator for both atherosclerosis and cancer.

Treatment and Prevention of Diseases of Aging

(1) Diet

Dietary fats and oils are best limited to 35 percent or less of total calories.[101,102] Consumption of fats and lipids that have been processed, exposed to air, heated, hydrogenated, or otherwise altered should be avoided when practical. Consumption of refined carbohydrates that are depleted of trace elements (white flour, white rice, and sugar) should be minimized. Total caloric intake should be moderated to maintain weight within 20 percent of ideal body weight. The use of excessive salt should be avoided. Diets should contain ample amounts of fiber-rich whole grains and fresh vegetables. Patients suffering with chronic debilitating diseases must be stricter with diet. Clinical improvement involves a healing process, often requiring months or years to complete.

(2) Nutritional Supplements

A scientifically balanced regimen of supplemental nutrients reinforces endogenous antioxidant defenses. It is not possible to receive optimal quantities of those substances from food alone. Supplemental antioxidants and vitamins should include vitamins E, C, B-1, B-2, B-3, B-6, B-12, folate, pantothenate, PABA, beta-carotene, coenzyme Q-10, and N-acetyl cysteine, plus a spectrum of minerals and trace elements including magnesium, zinc, copper, selenium, manganese, chromium, boron, and vanadium.[170] Trace elements can be toxic if taken to excess, and iron supplementation in the absence of deficiency will speed free radical damage. Iron should be supplemented only to treat deficiency states, confirmed by low serum ferritin and transferrin saturation.[271,296] Trace element supplementation should be under the supervision of a health care professional knowledgeable in nutrition. Dietary histories and biochemical testing allow supplementation to be tailored to the needs of each individual.[170,241-266,296,297]

(3) Modification of Health-Destroying Habits

Tobacco: It is best to eliminate the use of tobacco altogether, but, if that is not possible, a marked reduction in exposure would be helpful. This applies to cigarettes, pipe tobacco, cigars, snuff, and chewing tobacco. Tobacco causes problems, even without combustion. Free radical precursors are absorbed from tobacco through the lining of the mouth and nose, even without inhaling smoke. A relatively healthy adult with supplemental intake of antioxidants may tolerate a small exposure to tobacco without an increased risk of cancer, but even a small amount increases the risk of atherosclerosis.[298]

Alcohol: Many victims of degenerative diseases discover for themselves that alcohol is not well tolerated. For individuals with chronic illness, complete avoidance is advisable. A healthy adult should be able to tolerate and detoxify one to two ounces of pure ethanol per twenty-four hours (up to four eight-ounce glasses of beer, four small glasses of wine, or two shot glasses of hard liquor at most). That amount may be consumed in twenty-four hours without exceeding a healthy person's capacity to metabolize alcohol and neutralize the resulting free radicals.[101,102] But with daily use, tolerance may be lost.

(4) Physical Exercise

Moderate physical exercise, even a brisk forty-five minute walk several times per week will improve efficient utilization of oxygen. More vigorous aerobic exercise results in proportionately greater benefits. Lactate accumulates up to twice normal levels in tissues during endurance exercise.[299] Lactate has proven chelating properties, and it is possible that some of the benefits of exercise may result from chelating effects of lactate.[37]

(5) EDTA Chelation Therapy

The use of EDTA to alter the balance and distribution of essential metallic elements, while at the same time removing toxic heavy metals and catalytic free iron, has been shown to slow or arrest progression of many diseases of aging. Other benefits of chelation occur from uncoupling of disulfide and metallic cross-linkages between molecules, by normalization of calcium metabolism, by reactivation of enzymes poisoned by lead and other toxic metals, and by restoration of normal prostacyclin production along blood vessel walls. Lasting benefits follow a series of intravenous EDTA infusions, plus nutritional supplementation and lifestyle improvements.

This well-documented, safe, and effective therapy deserves widespread recognition and acceptance.

References

1. Clarke NE, Clarke CN, Mosher RE. The "in vivo" dissolution of metastatic calcium: an approach to atherosclerosis. *Am J Med Sci* 1955;229:142-149.

2. Schroeder HA, Perry HM, Jr. Antihypertensive effects of metal binding agents. *J Lab Clin Med* 1955;46:416.

3. Clarke NE, Clarke CN, Mosher RE. Treatment of angina pectoris with disodium ethylene diamine tetraacetic acid. *Am J Med Sci* 1956;232:654-666.

4. Boyle AJ, Casper JJ, McCormick H, et al. Studies in human and induced atherosclerosis employing EDTA. (Swiss, Basel) *Bull Schweiz Akad Med Wiss* 1957;13:408.

5. Muller SA, Brunsting LA, Winkelmann RK. Treatment of scleroderma with a new chelating agent, edathamil. *Arch Dermatol* 1959;80:101.

6. Clarke NE, Sr. Atherosclerosis, occlusive vascular disease and EDTA. *Am J Cardiol* 1960;6:233-236.

7. Meltzer LE, Ural ME, Kitchell JR. The treatment of coronary artery disease with disodium EDTA. In: Seven MJ, Johnson LA, eds. *Metal Binding in Medicine* Philadelphia: J. B. Lippincott Co; 1960:132-136.

8. Peters HA. Chelation therapy in acute, chronic and mixed porphyria. In: Seven MJ, Johnson LA, eds. *Metal Binding in Medicine.* Philadelphia: J. B. Lippincott Co; 1960:190-199.

9. Seven MJ, Johnson LA, eds. *Metal Binding in Medicine.* Proceedings of a Symposium Sponsored by Hahnemann Medical College and Hospital, Philadelphia. Philadelphia: J. B. Lippincott Co; 1960.

10. Kitchell JR, Meltzer LE, Seven MJ. Potential uses of chelation methods in the treatment of cardiovascular diseases. *Prog Cardiovasc Dis.* 1961;3:338-349.

11. Peters HA. Trace minerals, chelating agents and the porphyrias. *Fed Proc* 1961;3(2nd pt)(suppl 10):227-234.

12. Boyle AJ, Clarke NE, Mosher RE, et al. Chelation therapy in circulatory and sclerosing diseases. *Fed Proc* 1961;20(3)(2nd pt)(suppl 10):243-257.

13. Soffer A, Toribara T, Sayman A. Myocardial responses to chelation. *Br Heart J* 1961 Nov;23:690.

14. Peripheral Flow Opened Up. *Medical World News* Mar 15, 1963;4:36-39.

15. Boyle AJ, Mosher RE, McCann DS. Some in vivo effects of chelation-I: rheumatoid arthritis. *J Chronic Dis* 1963;16:325-328.

16. Aronov DM. First experience with the treatment of atherosclerosis patients with calcinosis of the arteries with trilon-B (disodium salt of EDTA). *Klin Med* (Russ Moscow) 1963;41:19-23.

17. Soffer A, Chenoweth M, Eichhorn G, et al. *Chelation Therapy* Springfield: Charles C. Thomas; 1964.

18. Soffer A. Chelation therapy for cardiovascular disease. In: Soffer A, ed. *Chelation Therapy* Springfield: Charles C. Thomas; 1964:15-33.

19. Lamar CP. Chelation therapy for occlusive arteriosclerosis in diabetic patients. *Angiology* 1964;15:379-394.

20. Friedel W, Schulz FH, Schoder L. Therapy of atherosclerosis through mucopolysaccarides and EDTA (ethylene diamine tetraacetic acid). (German) *Deutsch Gesundh* 1965;20:1566-1570.

21. Lamar CP. Chelation endarterectomy for occlusive atherosclerosis. *J Am Geriatr Soc* 1966;14:272-293.

22. Birk RE, Rupe CE. The treatment of systemic sclerosis with EDTA, pyridoxine and reserpine. *Henry Ford Hospital Medical Bulletin* 1966 June;14:109-139.

23. Lamar CP. *Calcium chelation of atherosclerosis, nine years' clinical experience.* Read before the Fourteenth Annual Meeting of the American College of *Angiology* San Juan, PR, Dec 8, 1968.

24. Olwin JH, Koppel JL. Reduction of elevated plasma lipid levels in atherosclerosis following EDTA chelation therapy. *Proc Soc Exp Biol Med* 1968;128:1137-1139.

25. Leipzig LJ, Boyle AJ, McCann DS. Case histories of rheumatoid arthritis treated with sodium or magnesium EDTA. *J Chronic Dis* 1970;22:553-563.

26. Brucknerova O, Tulacek J. Chelates in the treatment of occlusive atherosclerosis. (Czechoslavakian, Praha) *Vnitr Lek* 1972;18:729-735.

27. Nikitina EK, Abramova MA. Treatment of atherosclerosis patients with Trilon-B (EDTA). (Russian, Moscow) *Kardiologiia* 1972;12:137-139.

28. Evers R. Chelation of vascular atheromatous disease. *Journal International Academy Metabology* 1972;2:51-53.

29. Kurliandchikov VN. Treatment of patients with coronary arteriosclerosis with unithiol in combination with vitamins. (Russian, Kiev) *Vrach Delo* 1973;6:8.

30. Zapadnick VI, et al. Pharmacological activity of unithiol and its use in clinical practice. (Russian, Kiev) *Vrach Delo* 1973;8:122.

31. David O, Hoffman SP, Sverd J, et al. Lead and hyperactivity, behavioral response to chelation: a pilot study. *Am J Psychiatry* 1976;133:1155-1158.

32. Gordon GB, Vance RB. EDTA chelation therapy for atherosclerosis: history and mechanisms of action. *Osteopathic Annals* 1976;4:38-62.

33. Proceedings: Hearing on EDTA Chelation Therapy of the Ad Hoc Scientific Advisory Panel on Internal Medicine of the Scientific Board of the California Medical Society, March 26, 1976, San Francisco, California.

34. Halstead BW. *The Scientific Basis of EDTA Chelation Therapy* Colton, CA: Golden Quill Publishers; 1979.

35. Grumbles LA. *Radionuclide Studies of Cerebral and Cardiac Circulation Before and After Chelation Therapy.* Presented at a meeting of the American Academy of Medical Preventics, Chicago, Illinois, May 27, 1979.

36. Bjorksten J. The cross-linkage theory of aging as a predictive indicator. *Journal of Advancement in Medicine* 1989;2(1&2):59-76.

37. Bjorksten J. Possibilities and limitations of chelation as a means for life extension. *Journal of Advancement in Medicine* 1989;2(1&2):77-88.

38. Blumer W, Cranton EM. Ninety percent reduction in cancer mortality after chelation therapy with EDTA. *Journal of Advancement in Medicine* 1989;2(1&2):183-188.

39. Carpenter DG. Correction of biological aging. *Rejuvenation* 1980;8:31-49

40. Casdorph HR. EDTA chelation therapy, efficacy in arteriosclerotic heart disease. *Journal of Advancement in Medicine* 1989;2(1&2):121-130.

41. Casdorph HR. EDTA chelation therapy II, efficacy in brain disorders. *Journal of Advancement in Medicine* 1989;2(1&2):131-154.

42. McDonagh EW, Rudolph CJ, Cheraskin E. An oculocerebrovasculometric analysis of the improvement in arterial stenosis following EDTA chelation therapy. *Journal of Advancement in Medicine* 1989;2(1&2):121-130.

43. Olwin JH. *EDTA Chelation Therapy.* Presented at a meeting of the American Holistic Medical Association, University of Wisconsin, La Crosse, Wisconsin, May 28, 1981.

44. McDonagh EW, Rudolph CJ, Cheraskin E. The influence of EDTA salts plus multi-vitamin-trace mineral therapy upon total serum cholesterol/high-density lipoprotein cholesterol. *Medical Hypothesis* 1982;9:643-646.

45. McDonagh EW, Rudolph CJ, Cheraskin E. The effect of intravenous disodium ethylenediaminetetraacetic acid (EDTA) upon blood cholesterol in a private practice environment. *Journal of the International Academy of Preventive Medicine* 1982;7:5-12.

46. Williams DR, Halstead BW. Chelating agents in medicine. *J Toxicol: Clin Toxicol* 1983;19(10):1081-1115.

47. Casodorph HR, Farr CH. EDTA chelation therapy: treatment of peripheral arterial occlusion, an alternative to amputation. *Journal of Advancement in Medicine* 1989;2(1&2):167-182.

48. McDonagh EW, Rudolph CJ, Cheraskin E. The effect of EDTA chelation therapy with multivitamin/trace mineral supplementation upon reported fatigue. *Journal of Orthomolecular Psychiatry* 1984;13(4):277-279.

49. Riordan HD, Jackson JA, Cheraskin E. The effects of intravenous EDTA infusion on the multichemical profile. *American Clinical Laboratory* Oct 1988.

50. Riordan HD, Cheraskin E, Dirks M, et al: Electrocardiographic changes associated with EDTA chelation therapy. *Journal of Advancement in Medicine* 1988;1(4):191-194.

51. Deucher GP. EDTA chelation therapy: an antioxidant strategy. *Journal of Advancement in Medicine* 1988;1(4):182-190.

52. Rudolph CJ, McDonagh EW, Wussow DG. The effect of intravenous disodium ethylenediaminetetaacetic acid (EDTA) upon bone density. *Journal of Advancement in Medicine* 1988;1(2):79-85.

53. McDonagh EW, Rudolph DO, Cheraskin E. Effect of chelation therapy plus multivitamin/mineral supplementation upon vascular dynamics: ankle/brachial doppler blood pressure ratio. *Journal of Advancement in Medicine* 1989;2(1&2):159-166.

54. McDonagh EW, Rudolph DO, Cheraskin E. The "clinical change" in patients treated with EDTA chelation plus multivitamin/mineral supplemention. *Journal of Advancement in Medicine* 1989;2(1&2):189-196.

55. Olzewer E, Carter JP. EDTA chelation therapy: a retrospective study of 2,870 patients. *Journal of Advancement in Medicine* 1989;2(1&2):197-213.

56. Rudolph CJ, McDonagh EW, Barber RK. A non-surgical approach to obstructive carotid stenosis using EDTA chelation. *Journal of Advancement in Medicine* 1991;4(3):157-165.

57. Chappell LT, Stahl JP. The correlation between EDTA chelation therapy and improvement in cardiovascular function: a meta-analysis. *Journal of Advancement in Medicine* 1993;6(3):139-160.

58. Chappell LT, Stahl JP, Evans R. EDTA chelation treatment for vascular disease: a meta-analysis using unpublished data. *Journal of Advancement in Medicine* 1994;7(3):131-142.

59. Hancke C, Flytlie K. Benefits of EDTA chelation therapy on atherosclerosis: a retrospective study of 470 patients. *Journal of Advancement in Medicine* 1993;6(3):61-171.

60. Rudolph CJ, Samuels RT, McDonagh EW. Visual field evidence of macular degeneration reversal using a combination of EDTA

chelation and multiple vitamin and trace mineral therapy. *Journal of Advancement in Medicine* 1994;7(4):203-212.

61. Holliday HJ. Carotid restenosis: a case for EDTA chelation. *Journal of Advancement in Medicine* 1996;9(2):95-100.

62. Olwin JH, Kannabrocki EL, Sothern RB. Rationale for the use of EDTA-magnesium-heparin therapy in subjects with coronary artery disease. *Journal of Advancement in Medicine* 1997;10(2):157-165.

63. Olszewer E, Sabbag FC, Carter JP. A pilot double-blind study of sodium-magnesium EDTA in peripheral vascular disease. *J Natl Med Assn* 1990;82(3):174-177.

64. Doolan PD, Schwartz SL, Hayes JR, et al. An evaluation of the nephrotoxicity of ethylenediaminetetraacetate and diethylene-triaminepentaacetate in the rat. *Toxicol Appl Pharmacol* 1967;10:481-500.

65. Ahrens FA, Aronson AL: A comparative study of the toxic effects of calcium and chromium chelates of ethylenediaminetetraacetate in the dog. *Toxicol Appl Pharmacol* 1971;18:10-25.

66. Feldman EB: EDTA and angina pectoris. *Drug Therapy* 1975 Mar:62.

67. Wedeen RP, Mallik DK, Batuman V. Detection and treatment of occupational lead nephropathy. *Arch Intern Med* 1979,139:53-57.

68. Moel DI, Kuman K. Reversible nephrotoxic reactions to a combined 2, 3-dimercapto-1-propanol and calcium disodium ethylenediaminetetraacetic acid regimen in asymptomatic children with elevated blood levels. *Pediatrics* 1982;70(2):259-262.

69. McDonagh EW, Rudolph CJ, Cheraskin E. The effect of EDTA chelation therapy plus supportive multivitamin-trace mineral supplementation upon renal function: a study in serum creatinine. *Journal of Advancement in Medicine* 1989,2(1&2):235-243.

70. Cranton EM. Kidney effects of ethylene diamine tetraacetic acid (EDTA): a literature review. *Journal of Advancement in Medicine* 1989;2(1&2):227-233.

71. Batuman V, Landy E, Maesaka JK, et al. Contribution of lead to hypertension with renal impairment. *N Engl J Med* 1983;309(1):17-21.

72. McDonagh EW, Rudolph CJ, Cheraskin E. The effect of EDTA chelation therapy plus supportive multivitamin-trace mineral supplementation upon renal function: a study in blood urea nitrogen (BUN), *J Holistic Med* 1989;2(1&2):251-261.

73. Sehnert KW, Claque AF, and Cheraskin E. The improvement in renal function following EDTA chelation and multivitamin trace mineral therapy: a study in creatinine clearance. *Journal of Advancement in Medicine* 1989;2(1&2):245-250.

74. Riordon HD, Cheraskin E, Dirks M, et al. Another look at renal function and the EDTA treatment process. *Journal of Advancement in Medicine* 1989;2(1&2):263-268.

75. Kitchell JR, Palmon F, Aytan N, et al. The treatment of coronary artery disease with disodium EDTA, a reappraisal. *Am J Cardiol* 1963;11:501-506.

76. Cranton EM, Frackelton JP. Current status of EDTA chelation therapy in occlusive arterial disease. *Journal of Advancement in Medicine* 1989;2(1&2):107-119.

77. Wartman A, Lampe TL, McCann DS, et al. Plaque reversal with MgEDTA in experimental atherosclerosis: elastin and collagen metabolism. *J Atheros Res* 1967;7:331.

78. Wissler RW: Principles of the pathogenesis of atherosclerosis. In: Braunwald E, ed. *Heart Disease* Philadelphia: W. B. Saunders Co; 1980:1221-1236.

79. Kjeldsen K, Astrup P, Wanstrup J. Reversal of rabbit atherosclerosis by hyperoxia. *J Atheros Res* 1969;10:173.

80. Vesselinovitch D, Wissler RW, Fischer-Dzoga K, et al. Regression of atherosclerosis in rabbits. I. Treatment with low fat diet, hyperoxia and hypolipidemic agents. *Atherosclerosis* 1974;19:259.

81. Sincock A. Life extension in the rotifer by application of chelating agents. *J Gerontot* 1975;30:289-293.

82. Wissler RW, Vesselinovitch D. Regression of atherosclerosis in experimental animals and man. *Mod Concepts Cardiovasc Dis* 1977,46:28.

83. Walker F. *The Effects of EDTA Chelation Therapy on Plaque, Calcium, and Mineral Metabolism in Arteriosclerotic Rabbits.* Ph.D. thesis. Texas State University, 1980. (Available from University Microfilm International, Ann Arbor, MI 48016.)

84. Guldager B, Jelnes R, Jorgensen SJ, et al. EDTA treatment of intermittent claudication—a double-blind, placebo-controlled study. *Journal of Internal Medicine* 1992;231:261-267.

85. Sloth-Nielsen J, Guldager B. Mouritzen C. Arteriographic findings in EDTA chelation therapy on peripheral atherosclerosis. *The American Journal of Surgery.* 1991;162:122-125.

86. Van Rij AM, Solomon C, Packer SG. Chelation therapy for intermittent claudication. A double-blind, randomized, controlled study. *Circulation.* 1994 Sep;90(3):1194-1199.

87. Diehm C, Wilhelm C, Poeschl J. *Effects of EDTA-Chelation Therapy in Patients with Peripheral Vascular Disease—A Double-Blind Study.* An unpublished study performed by the Department of Internal Medicine, University of Heidelberg, Heidelberg, Germany in 1985. Presented as a paper before the International Symposium of Atherosclerosis, Melbourne, Australia, October 14, 1985.

88. Carter JP. If EDTA chelation therapy is so good, why is it not more widely accepted? *Journal of Advancement in Medicine* 1989;2(1&2):213-226.

89. Chappell LT. Disputes author's conclusions on effectiveness of EDTA chelation therapy. *Alternative Therapies.* Sep 1996;2(5):16-17.

90. Fossel M. Role of cell senescence in human aging. *Journal of Anti-Aging Medicine* 2000;3(1):91-98.

91. Fossel M. Telomerase and the aging cell: implications for human health. *JAMA* 1998;279:1732-1735.

92. Fossel M. *Reversing Human Aging.* New York: William Morrow and Company; 1996.

93. Benchimol S, Vaziri H. Reconstruction of telomerase activity in normal cells leads to elongation of telomeres and extended replicative life span. *Curr Biol* 1998;8:279-282.

94. Chang E, Harley CB. Telomere length as a measure of replicative histories in human vascular tissues. *Proc Natl Acad Sci USA* 1995;92:1190-1194

95. Frustaci A, Magnavita N, Chirmenti C, et al. Marked elevation of myocardial trace elements in idiopathic dilated cardiomyopathy compared with secondary cardiac dysfunction. *JACC* 1999;33(6):1578-1583.

96. Tatton WG, Olanow CW. Apoptosis in neurodegenerative diseases: the role of mitochondria. *Biochem Biophy Acta* 1999;1410(2):195-213 (a review).

97. Peng L, Nijjhawan D, Budihardya SM, et al. Cytochrome c- and d-ATP-dependent Apaf-l/caspase-9 complex initiate an apoptotic protease cascade. *Cell* 1997;91:479-489.

98. Demopoulos HB, Pietronigro DD, Flamm ES, et al. The possible role of free radical reactions in carcinogenesis. *Journal of Environmental Pathology and Toxicology* 1980;3:273-303.

99. Harman D. The aging process. *Proc Natl Acad Sci USA* 1981;78:7124-7128.

100. Dormandy TL. An approach to free radicals. *Lancet* 1983,ii.1010-14.

101. Demopoulos HB, Pietronigro DD, Seligman ML. The development of secondary pathology with free radical reactions as a threshold mechanism. *Journal of the American College of Toxicology* 1983;2(3):173-184.

102. Demopoulos HB. *Molecular Oxygen in Health and Disease.* Read before the American Academy of Medical Preventics Tenth Annual Spring Conference, Los Angeles, California May 21, 1983.(Available on three audio cassettes from Instatape, P.O. Box 1729, Monrovia, CA 91016.)

103. Ames BN. Dietary carcinogens and anticarcinogens. *Science.* 1983;221:1256-1264.

104. Dormandy TL. Free-radical reaction in biological systems. *Ann R Coll Surg Engl* 1980;62:188-194.

105. Dormandy TL. Free-radical oxidation and antioxidants. *Lancet* 1978;8:647-650.

106. Levine SA, Reinhardt JH. Biochemical-pathology initiated by free radicals, oxidant chemicals, and therapeutic drugs in the etiology of chemical hypersensitivity disease. *Journal of Orthomolecular Psychiatry* 1983;12(3):166-183.

107. Del Maestro RF. An approach to free radicals in medicine and biology. *Acta Physiol Scand* 1980;492(supp1):153-68.

108. Poole CP. *Electron Spin Resonance A Comprehensive Treatise on Experimental Techniques.* New York: Interscience Publishers; 1967.

109. Demopoulos HB, Flamm ES, Seligman ML, et al. Membrane perturbations in central nervous system injury: theoretical basis for free radical damage and a review of the experimental data. In: Popp AJ, Bourke LR, Nelson LR, Kimelbert HK, eds. *Neural Trauma.* New York: Raven Press; 1979:63-78.

110. Seligman ML, Mitamura JA, Shera N, et al. Corticosteroid (methylprednisolone) modulation of photoperoxidation by ultraviolet light in liposomes. *Photochem Photobiol* 1979;29:549-558.

111. Pryor WA. Free radical reactions and their importance in biochemical systems. *Fed Proc* 1973;32:1862-1869.

112. Pryor WA, ed. *Free Radicals in Biology Volumes 1-3.* New York: Academic Press; 1976.

113. Lambert L, Willis ED. The effect of dietary lipid peroxides, sterols and oxidized sterols on cytochrome P450 and oxidative demethylation. *Biochem Pharmacol* 1977a;26:1417-1421.

114. Lambert C, Willis ED. The effect of dietary lipids on 3, 4 benzo(a)pyrene metabolism in the hepatic endoplasmic reticulum. *Biochem Pharmacol* 1977b;26:1423-1477.

115. Fedorenko VI. Effect of cysteine, glutathione and 1-p-chlorophenyltetrazole-thione-2 on postradiation changes in the metabolic free radical content of albino rat tissues. *Radiobiologiia* 1979;19:67-73.

116. Fridovich I. Superoxide dismutases. *Annu Rev Biochem* 1975:147-159.

117. Black HS, Chan JT. Experimental ultraviolet light-carcinogenesis. *Photochem Photobiol* 1977;26:183-189.

118. Eaton GJ, Custer P, Crane R. Effects of ultraviolet light on nude mice: cutaneous carcinogenesis and possible leukemogenesis. *Cancer* 1978;42:182-188.

119. Tappel AL. Lipid peroxidation damage to cell components. *Fed Proc* 1973;32:1870-1874.

120. Kotin P, Falk HL. Organic peroxide, hydrogen peroxide, epoxides and neoplasia. *Radiat Res* 1963;3(suppl):193-211.

121. Demopoulos, HB. Control of free radicals in the biologic systems. *Fed Proc.* 1973;32:1903-1908.

122. Walling C. Forty years of free radicals. In: Pryor WA, ed. *Organic Free Radicals* Washington, DC: American Chemical Society; 1978.

123. Demopoulos HB. The basis of free radical pathology. *Fed Proc* 1973;32:1859-1861.

124. Tappel AL. Will antioxidant nutrients slow aging processes? *Geriatrics* 1968;23:97-105.

125. Coon MJ. Oxygen activation in the metabolism of lipids, drugs and carcinogens. *Nutr Rev* 1978;36:319-328.

126. Coon MJ. Reconstitution of the cytochrome P-450-containing mixed-function oxidase system of liver microsomes. *Methods Enzymol* 1978;52:200-206.

127. Coon MJ, van der Hoeven TA, Dahl SB, et al. Two forms of liver microsomal cytochrome P-450, P-4501m2 and P-450M4 (rabbit liver). *Methods Enzymol* 1978;52:109-117.

128. Panganamala RV, Sharma HM, Sprecher H, et al. A suggested role for hydrogen peroxide in the biosynthesis of prostaglandins. *Prostaglandins* 1974;8:3-11.

129. Maisin JR, Decleve A, Gerber GB, et al. Chemical protection against the long-term effects of a single whole-body exposure of mice to ionizing radiation. II. Causes of death. *Radiat Res* 1978;74:415-435.

130. McGinnes JE, Proctor PH, Demopoulos HB, et al. In vivo evidence for superoxide and peroxide production by adriamycin and cis-platium. In: Autor A, ed. *Active Oxygen and Medicine* New York: Raven Press; 1980.

131. Petkau A. Radiation protection by superoxide dismutase. *Photochem Photobiol* 1978;28:765-774.

132. Schaefer A, Komlos M, Seregi A. Lipid peroxidation as the cause of the ascorbic acid induced decrease of ATPase activities of rat brain microsomes and its inhibition by biogenic amines and psychotropic drugs. *Biochem Pharmcol* 1975;24: 1781-1786.

133. Vladimirov YA, Sergeer PV, Seifulla RD, et al. Effect of steroids on lipid peroxidation and liver mitochondrial membranes. *Molekuliarnaia Biologiia* (Russian, Moscow) 1973;7:247-262. (Translated by Consultants Bureau, a division of Plenum Publishing Inc., New York.)

134. Sies H, Summer KH. Hydroperoxide-metabilizing systems in rat liver. *Eur J Biochem* 1975;57:503-512.

135. Alfthan G, Pikkarainen J, Huttunen JK, et al. Association between cardiovascular death and myocardial infarction and serum selenium in a matched-pair longitudinal study. *Lancet* 1982;2(8291):175-179.

136. Willett WC, Morris JS, Pressel S, et al. Prediagnostic serum selenium and risk of cancer. *Lancet* 1983;2(8343):130-134.

137. Demopoulos HB, Flamm ES, Seligman ML, et al. Antioxidant effects of barbiturates in model membranes undergoing free radical damage. *Acta Neurol Scand* 1977;56(suppl 64):152.

138. Flamm ES, Demopoulos HB, Seligman ML, et al. Barbiturates and free radicals. In: Popp AJ, Popp RS, Bourke LR, Nelson HB, Kimelberg HK, eds. *Neural Trauma*. New York: Raven Press; 1979:289-300.

139. Butterfield JD, McGraw CP. Free radical pathology. *Stroke* 1978;9(5):443-445.

140. Demopoulos HB, Flamm ES, Pietronigro DD, et al. The free radical pathology and the microcirculation in the major

central nervous system disorders. *Acta Physiol Scand* 1980;492(suppl):91-119.

141. Gey KF, Puska P, Jordan P, et al. Inverse correlation between plasma vitamin E and mortality from ischemic heart disease in cross-cultural epidemiology. *Am J Clin Nutr* 1991;53:326S-334S.

142. Seubens WE, Smith S. Serum cholesterol correlations with atherosclerosis at autopsy. *American Clinical Laboratory* 1997 Apr;14-15.

143. Morel DW, Hessler JR, Chisolm GM. Low-density lipoprotein cytotoxicity induced by free radical peroxidation of lipid. *Journal of Lipid Research* 1983;24:1070-1076.

144. Smith LL. *Cholesterol Autooxidation*. New York: Plenum Press; 1981.

145. Gaby AR. Nutritional factors in cardiovascular disease. *Journal of Advancement in Medicine* 1989;2(1&2):89-105.

146. Taylor CB, Peng SK, Werthessen NT, et al. Spontaneously occurring angiotoxic derivatives of cholesterol. *Am J Clin Nutr* 1979;32:40.

147. Cortese C, Bernardini S, Motti C. Atherosclerosis in light of the evidence from large statin trials. *Ann Ital Med Int* 2000;15(1):103-107

148. Song YM, Sung J, Kim JS. Which cholesterol level is related to the lowest mortality in a population with low mean cholesterol level: a 6.4-year follow-up study of 482,472 Korean men. *Am J Epidemiol* 2000;151(8):739-747.

149. Wilcken DEL, Wilcken B. The pathogenesis of coronary artery disease: a possible role for methionine metabolism. *J Clin Investig* 1976;57:1079-1082.

150. Boushey CJ, Beresford SAA, Omenn GS, et al. A quantitative assessment of plasma homocysteine as a risk factor for vascular disease: probable benefits of increasing folic acid intakes. *JAMA* 1995;274:1049-1057.

151. Clarke R, Daly LE, Robinson K, et al. Hyperhomocysteinernia: an independent risk factor for vascular disease. *N Engl J Med* 1991;324:1149-1155.

152. Graham IM, Daly LE, Refsum HM, et al. Plasma homocysteine as a risk factor for vascular disease: the European concerted action project. *JAMA* 1997;277:1775-1781.

153. Nygard O, Vollsett SE, Refsum HM, et al. Total plasma homocysteine and cardiovascular risk profile: the Hordaland hornocysteine study. *JAMA* 1995;274:1526-1533.

154. Clopath P, Smith VC, McCully KS. Growth promotion by homocysteic acid. *Science* 1976;192:372-374.

155. Tsai J-C, Perella MA, Yoshizumi M, et al. Promotion of vascular smooth muscle cell growth by homocysteine: a link to atherosclerosis. *Proc Natl Acad Sci USA* 1994;91:6369-6373.

156. Welch GN, Upchurch GR, Loscalzo J. Homocysteine, oxidative stress, and vascular disease. *Hosp Pract* 1997;32:81-92.

157. D'Angelo A, Selhub J. Homocysteine and thrombotie disease. *Blood* 1997;90:1-11.

158. Naruszewicz M, Mirkiewicz E, Olszewski AJ, et al. Thiolation of low-density lipoprotein by homocysteine thiolactone causes increased aggregation and altered interaction with cultured macrophages. *Nutr Metab Cardiovasc Dis* 1994;4:70-77.

159. McCully KS. Homocysteine and vascular disease. *Nat Med* 1996;2:386-389.

160. McCully KS. Homocysteine, folate, vitamin B-6, and cardiovascular disease. Editorial. *JAMA* 1998;279(5):392-393.

161. McCully KS. Vascular pathology of hornocysteinernia: implications for the pathogenesis of arteriosclerosis. *Am J Pathol* 1969;56:111-128.

162. Rimm EB, Willett WC, Hu FB, et al. Folate and vitamin B-6 from diet and supplements in relation to risk of coronary heart disease among women. *JAMA* 1998;279(5):359-364.

163. Boushey CJ, Beresford SAA, Omenn GS, et al. A quantitative assessment of plasma homocysteine as a risk factor for vascular disease: probable benefits of increasing folic acid intakes. *JAMA* 1996;274:1049-1057.

164. Brattstrom L, Israelsson B, Norrving B, et al. Impaired homocysteine metabolism in early onset cerebral and peripheral occlusive arterial disease: effects of pyridoxine and folic acid treatment. *Atherosclerosis* 1990;81:51-60.

165. Jacob RA, Wu M, Henning SM, et al. Homocysteine increases as folate decreases in plasma of healthy men during short-term dietary folate and methyl group restriction. *J Nutr* 1994;124:1072-1080.

166. Naurath HJ, Joosten E, Riezler R, et al. Effects of vitamin B-12, folate, and vitamin B-6 supplements in elderly people with normal serum vitamin concentrations. *Lancet* 1996,346:85-89.

167. Selhub J, Jacques PF, Bostom AG, et al. Association between plasma homocysteine concentrations and extracranial carotid-artery stenosis. *N Engl J Med* 1995;332:289-291.

168. Nygard O, Nordrehaug JE, Refsurn H, et al. Plasma homocysteine levels and mortality in patients with coronary artery disease. *N Engl J Med* 1997;337:230-236.

169. Rimin EB, Stampfer MJ, Ascherio A, et al. Dietary folate, vitamin B-6, and vitamin B-12 intake and risk of CHD among a large population of men. Abstract. *Circulation* 1996;93:625.

170. Passwater RA, Cranton EM. *Trace Elements, Hair Analysis and Nutrition*. New Canaan, Ct: Keats Publishing, Inc; 1983.

171. Fourcans B. Role of phospholipids in transport and enzymic reactions. *Adu Lipid Res* 1974:147-226.

172. Babior BM. Oxygen-dependent microbial killing by phagocytes. *N Engl J Med* 1978;298:659-668.

173. Rosen H. Klebanoff SJ. Bactericidal activity of a superoxide anion-generating system. *J Exp Med* 1979;149:27-39.

174. Masterson WL, Slowinski E. In: *Chemical Principles*. Philadelphia: Saunders; 1977:203(plate 5).

175. Mayes PA. Biologic oxidation. In: Martin DW, Mayes PA, Rodwell VW, eds. *Harper's Review of Biochemistry*. Los Alton, Ca: Lange Medical Publications; 1983:129-130.

176. March J. *Advanced Organic Chemistry Reactions, Mechanics, and Structure.* 2nd ed. New York: McGraw-Hill; 1978:620.

177. Flamm ES, Demopoulos HB, Seligman ML, et al. Free radicals in cerebral ischemia. *Stroke* 1978;9(5):445-447.

178. *Hyperbaric Oxygen Therapy: A Committee Report.* February 1981. Undersea Medical Society, Inc., 9650 Rockville Pike, Bethesda, MD 20014 (UMS Publication Number 30 CR(HBO) 2-23-81).

179. Foote CS. Chemistry of singlet oxygen VII. Quenching by beta carotene. *J Am Chem Soc* 1968;90:6233.

180. Rimm EB, Stampfer MJ, Ascheno A, et al. Vitamin E consumption and the risk of coronary heart disease in men. *N Engl J Med* 1993 May 20;328(20):1450-1456.

181. Stampher MJ, Hennekens CH, Morrison JE, et al. Vitamin E consumption and the risk of coronary disease in women. *N Engl J Med* 1993 May 20;328(20):1444-1449.

182. Rimm EB, Stampfer MJ, Ascheno A, et al. Vitamin E consumption and the risk of coronary heart disease in men. *N Engl J Med* 1993 May 20;328(20):1450-1456

183. Peto R, Doll R, Buckley JD, et al. Can dietary beta-carotene materially reduce human cancer rates? *Nature* 1981;290:201.

184. Stoyanovsky DA, et al. Endogenous ascorbate regenerates vitamin E in the retina directly and in combination with exogenous dihydrolipoic acid. *Curr Eye Res* 1995 Mar;14(3):181-189.

185. Chan AC. Partners in defense, vitamin E and vitamin C. *Can J Physiol Pharmacol* 1993 Sep;71(9):725-731.

186. Ho CT, et al. Regeneration of vitamin E in rat polymorphonuclear leucocytes. *FEBS Lett* 1992 Jul 20;306(2-3):269-272.

187. Chan AC, et al. Regeneration of vitamin E in human platelets. *J Biol Chem* 1991 Sep 15;266(26):17290-17295.

188. Tappel AL. Will antioxidant nutrients slow aging processes? *Geriatrics* 1968 Oct;23(10):97-105.

189. Enstrom EE, Kanim LE, Klein MA. Vitamin C intake and mortality among a sample of the United States population. *Epidemiology* 1992;3(3):194-202.

190. Schaefer A, Komlos M, Seregi A. Lipid peroxidation as the cause of the ascorbic acid induced decrease of ATPase activities of rat brain microsomes and its inhibition by biogenic amines and psychotropic drugs. *Biochem Pharmacol* 1975;24:1781-86.

191. Ito T, Allen N, Yashon D. A mitochondrial lesion in experimental spinal cord trauma. *J Neurosurg* 1978;48:434-442.

192. Gillman MW, Cupples LA, Gagnon D, et al. Margarine intake and subsequent coronary heart disease in men. *Epidemiology* 1997 Mar;8(2):144-149.

193. Atherosclerosis and auto-oxidation of cholesterol. *Lancet* 1980;ii:964-965.

194. Pietronigro DD, Demopoulos HB, Hovsepian M, et al. Brain ascorbic acid (AA) depletion during cerebral ischemia. *Stroke* 1982;13(1):117.

195. Demopoulos HB, Flamm ES, Seligman ML, et al. Further studies on free-radical pathology in the major central nervous system disorders: effect of very high doses of methylprednisolone on the functional outcome; morphology and chemistry of experimental spinal cord impact injury. *Can J Physiol Pharmacol* 1982;60(11):1415-1424.

196. Flamm ES, Demopoulos HB, Seligman ML, et al. Free radicals in cerebral ischemia. *Stroke* 1978;9:445.

197. Demopoulos HB, Flamm ES, Seligman ML, et al. Oxygen free radicals in central nervous system ischemia and trauma. In: Autor AP, ed. *Pathology of Oxygen* New York: Academic Press; 1982:127-155.

198. Demopoulos HB, Flamm ES, Seligman ML, et al. Molecular pathogenesis of spinal cord degeneration after traumatic injury. In: Naftchi NE, ed. *Spinal Cord Injury* New York and London: Spectrum Publications, Inc; 1982:45-64.

199. Sukoff MH, Hollin SA, Espinosa OK, et al. The protective effect of hyperbaric oxygenation in experimental cerebral edema. *J Neurosur* 1968;29:236-239.

200. Kelly DL Jr, Lassiter KRL, Vongsvivut A, et al. Effects of hyperbaric oxygen and tissue oxygen studies in experimental paraplegia. *J Neurosurg* 1972;36:425-429.

201. Holbach KH, Wassman H, Hoheluchter KL, et al. Clinical course of spinal lesions treated with hyperbaric oxygen. *Acta Neurochir* 1975;31:297-298.

202. Holbach KH, Wassman H, Linke D. The use of hyperbaric oxygenation in the treatment of spinal cord lesions. *Eur Neurol* 1977;16:213-221.

203. Yeo JD, Stabback S, McKinsey B. Study of the effects of hyperbaric oxygenation on experimental spinal cord injury. *Med J Aust* 1977;2:145-147.

204. Jones RF, Unsworth IP, Marasszeky JE. Hyperbaric oxygen and acute spinal cord injuries in humans. *Med J Aust* 1978;2:573-575.

205. Yeo JD, Lawry C. Preliminary report on ten patients with spinal cord injuries treated with hyperbaric oxygenation. *Med J Aust* 1978,2:572-573.

206. Gelderd JB, Welch DW, Fife WP, et al. Therapeutic effects of hyperbaric oxygen and dimethyl sulfoxide following spinal cord transections in rats. *Undersea Biomedical Research* 1980;7:305-320.

207. Sukoff MH. Central nervous system: review and update cerebral edema and spinal cord injuries. *HBO Review* 1980,1:189-195.

208. Jesus-Greenberg DA. Acute spinal cord injury and hyperbaric oxygen therapy: a new adjunct in management. *Journal of Neurosurgical Nursing* 1980:12:155-160.

209. Higgins AC, Pearlstein MS, Mullen JB, et al. Effects of hyperbaric oxygen therapy on long-tract neuronal conduction in the acute phase of spinal cord injury. *J Neurosurg* 1981;55(4):501-510.

210. Sukoff MH, Ragatz RE. Use of hyperbaric oxygen for acute cerebral edema. *Neurosurgery* 1982;10:29-38.

211. De La Torre JC, Johnson CM, Goode DJ, et al. Pharmacologic treatment and evaluation of permanent experimental spinal cord trauma. *Neurology* 1975;25:508-514.

212. De La Torre JC, Kawanaga HM, Rowed DW, et al. Dimethyl sulfoxide in central nervous system trauma. *Ann NY Acad Sci* 1975,243:362-389.

213. De La Torre JC, Surgeon JW. Dexamethasone and DMSO in experimental transorbital cerebral infarction. *Stroke* 1976,7:577-583.

214. Laha RK, Dujovny M, Barrionuevo PJ, et al. Protective effects of methyl prednisolone and dimethyl sulfoxide in experimental middle cerebral artery embolectomy. *J Neurosurg* 1978;49:508-516.

215. Fischer BH, Marks M, Reich T. Hyperbaric-oxygen treatment of multiple sclerosis. *N Eng J Med* 1983;308:181-186.

216. Swank R. *A Biochemical Basis of Multiple Sclerosis.* Springfield, Ill: Charles C. Thomas; 1961.

217. Cimino JA, Demopoulos HB. Introduction: determinants of cancer relevant to prevention, in the war on cancer. *Journal of Environmental Pathology and Toxicology* 1980;3:1-10.

218. Seligman ML, Flamm ES, Goldstein BD, et al. Spectrofluorescent detection of malonaldehyde as a measure of lipid free radical damage in response to ethanol potentiation of spinal cord trauma. *Lipids* 1977;12(11):945-950.

219. Flamm ES, Demopoulos HB, Seligman ML, et al. Ethanol potentiation of central nervous system trauma. *J Neurosurg* 1977;46:328-334.

220. Dix T. Metabolism of polycyclic aromatic hydrocarbon derivatives to ultimate carcinogens during lipid peroxidation. *Science* 1983;221:277.

221. Samuelsson B. Leukotrienes: mediators of immediate hypersensitivity reactions and inflammation. *Science* 1983;220:568-575

222. Hess ML, Manson NH, Okabe E. Involvement of free radicals in the pathophysiology of ischemic heart disease. *Can J Physiol Pharmacol* 1982;60(11):1382-1389.

223. Vincent GM, Anderson JL, Marshall HW. Coronary spasm producing coronary thrombosis and myocardial infarction. *N Engl J Med* 1983;309(14):220-239.

224. Harmon D. *The Free Radical Theory of Aging.* Read before the Orthomolecular Medical Society, San Francisco, California, May 8, 1983. (Available on audio cassette from AUDIO-STATS, 3221 Carter Avenue, Marina Del Rey, CA 90291.)

225. Singal PK, Kapur N, Dhillon KS, et al. Role of free radicals in catecholamine-induced cardiomyopathy. *Can J Physiol Pharmacol* 198260(11):1340-1397.

226. Crapper-McLaughlin DR. *Aluminum Toxicity in Senile Dementia: Implications for Treatment.* Read before the Fall Conference, American Academy of Medical Preventics, Las Vegas, Nevada, Nov 8, 1981.

227. Raymond JP, Merceron R, Isaac R, et al. *Effects of EDTA and Hypercalcemia on Plasma Prolactin, Parathyroid Hormone and Calcitonin in Normal and Parathyroidectomized Individuals.* Read before the Frances and Anthony D'Anna International Memorial Symposium, Clinical Disorders of Bone and Mineral Metabolism, May 8, 1983. (Abstract available from Henry Ford Hospital, Dearborn, Michigan.)

228. Frost HB. Coherence treatment of osteoporosis. *Orthop Clin North Am* 1981;12:649-669.

229. DeLuca HF, Frost HM, Jee WSS, et al, eds. *Osteoporosis: Recent Advances in Pathogenesis and Treatment.* Baltimore: University Press; 1981.

230. Frost HM. Treatment of osteoporosis by manipulation of coherent bone cell populations. *Clin Orthop* 1979;143:227-244.

231. Meyer MS, Chalmers TM, Reynolds JJ. *Inhibitory Effect of Follicular Stimulating Hormone in Parathormone in Rat Calvaria In Vitro.* Read before the Frances and Anthony D'Anna

International Memorial Symposium, Clinical Disorders of Bone and Mineral Metabolism, May 8, 1983. (Abstract available from Henry Ford Hospital, Dearborn, Michigan.)

232. Wills ED. Lipid peroxide formation in microsomes. *Biochem J* 1969;113:325-332.

233. Gutteridge JMC, Rowley DA, Halliwell B, et al. Increased non-protein-bound iron and decreased protection against superoxide-radical damage in cerebrospinal fluid from patients with neuronal ceroid lipofuscinoses. *Lancet* 1982;ii:459-460.

234. Willson RL. Iron, zinc, free radicals and oxygen tissue disorders and cancer control. In: *Iron Metabolism*. Ciba Foundn Symp 51 (new series). Amsterdam: Elsevier; 1977:331-354.

235. Gutteridge JMC. Fate of oxygen free radicals in extracellular fluid. *Biochem Soc Trans* 1982;10:72-74.

236. Wills ED. Mechanisms of lipid peroxide formation in tissues: role of metals and haematin proteins in the catalysis of the oxidation of unsaturated fatty acids. *Biochem Biophys Acta* 1965;98:238-251.

237. Gutteridge JMC, Rowley DA, Halliwell B. Superoxide-dependent formation of hydroxyl radicals and lipid peroxidation in the presence of iron salts. *Biochem J* 1982;206:605-609.

238. Heys AD, Dormandy TL. Lipid peroxidation in iron-overloaded spleens. *Clinical Science* 1981;60:295-301.

239. Ericson JE, Shirahata H, Patterson CC. Skeletal concentrations of lead in ancient Peruvians. *N Engl J Med* 1979;300:946-951.

240. Schroder HA. *The Poisons Around Us*. Bloomington: Indiana University Press; 1974:49.

241. Jenkins DW. *Toxic Trace Metals in Mammalian Hair and Nails*. US Environmental Protection Agency publication No.(EPA)-600/4-79049. Environmental Monitoring Systems Laboratory, 1979. (Available from National Technical Information Service, U.S. Department of Commerce, Springfield, VA 22161.)

242. Cranton EM, Bland JS, Chatt A, et al. Standardization and interpretation of human hair for elemental concentrations. *J Holistic Med* 1982;4:10-20.

243. Hansen JC, Christensen LB, Tarp U. Hair lead concentration in children with minimal cerebral dysfunction. *Danish Med Bull* 1980;27:259-262.

244. Medeiros DM, Pellum LK, Brown BJ. The association of selected hair minerals and anthropometric factors with blood pressure in a normotensive adult population. *Nutr Research* 1983;3:51-60.

245. Moser PB, Krebs NK, Blyler E. Zinc hair concentrations and estimated zinc intakes of functionally delayed normal sized and small-for-age children. *Nutr Research* 1982;2:585-590.

246. Thimaya S, Ganapathy SN. Selenium in human hair in relation to age, diet, pathological condition and serum levels. *Sci Total Environ* 1982;24:41-49.

247. Musa-Alzubaida L, Lombeck I, Kasperek K, et al. Hair selenium content during infancy and childhood. *Eur J Pediatr* 1982:139:295-296.

248. Gibson RS, Gage L. Changes in hair arsenic levels in breast and bottle fed infants during the first year of infancy. *Sci Total Environ* 1982;26:33-40.

249. Ely DL, Mostardi RA, Woebkenberg N, et al. Aerometric and hair trace metal content in learning-disabled children. *Environ Res* 1981;25(2):325-339.

250. Yokel RA. Hair as an indicator of excessive aluminum exposure. *Clin Chem* 1982;28(4):662-665.

251. Bhat RK, et al. Trace elements in hair and environmental exposure. *Sci Total Environ* 1982;22(2):169-178.

252. Hurry VJ, Gibson RS. The zinc, copper, and manganese status of children with malabsorption syndromes and inborn errors of metabolism. *Biol Trace Element Res* 1982;4:157-173.

253. Thatcher RW, Lester ML, McAlester R, et al. Effects of low levels of cadmium and lead on cognitive functioning in children. *Arch Environ Health* 1982;37(3):159-166.

254. Peters HA, Croft WA, Woolson EA, et al. Arsenic, chromium, and copper poisoning from burning treated wood. *N Engl J Med* 1983:308(22):1360-1361.

255. Yamanaka S, Tanaka H, Nishimura M. Exposure of Japanese dental workers to mercury. *Bull Toyko Den Coll* 1982;23:15-24.

256. Capel ID, Spencer EP, Levitt HN, et al. Assessment of zinc status by the zinc tolerance test in various groups of patients. *Clin Biochem* 1982;15(2):257-260.

257. Vanderhoof JA, et al. Hair and plasma zinc levels following exclusion of biliopancreatic secretions from functioning gastrointestinal tract in humans. *Dig Dis Sci* 1983;28(4):300-305.

258. Foli MR, Henningan C, Errera J. A comparison of five toxic metals among rural and urban children. *Environ Pollut Ser A Ecol Biol* 1982;29:261-270.

259. Collipp PJ, Kuo B, Castro-Magana M, et al. Hair zinc levels in infants. *Clin Pediatr* 1983;22(7):512-513.

260. Medeiros DM, Borgman RF. Blood pressure in young adults as associated with dietary habits, body conformation, and hair element concentrations. *Nutr Res* 1982;2:455-466.

261. Huel G. Boudene C, Ibrahim MA. Cadmium and lead content of maternal and newborn hair: relationship to parity, birth weight, and hypertension. *Arch Environ Health* 1981;35(5):221-227.

262. Marlowe M, Folio R, Hall D, Errera J. Increased lead burdens and trace mineral status in mentally retarded children. *J Spec Educ* 1982;16:87-99.

263. Marlowe M, Errera J, Stellern J, et al. Lead and mercury levels in emotionally disturbed children. *J Orthomol Psychiatr* 1983;12(4):260-267.

264. Nolan KR. Copper toxicity syndrome. *J Orthomol Psychiatr* 1983;12(4):270-282.

265. Klevay L. Hair as a biopsy material-assessment of copper nutriture. *Am J Clin Nutr* 1970;23(8):1194-1202.

266. Rees EL. Aluminum poisoning in Papua New Guinea natives as shown by hair testing. *J Orthomol Psychiatr* 1983;12(4):312-313.

267. Cook JD, Finch CA, Smith NJ. Evaluation of the iron status of a population. *Blood* 1976;48:449-455.

268. Kannel WB, Hjortland MC, McNamara PM, et al. Menopause and the risk of cardiovascular disease. The Framingham Study. *Ann Intern Med* 1976;85:447-452

269. Hjortland MC, McNamara PM, Kannel WB. Some atherogenic concomitants of menopause: The Framingham Study. *Am J Emidemiol* 1976;103:304-311.

270. Gordon T, Kannel WB, Hjortland MC, et al. Menopause and coronary heart disease. The Framingham Study. *Ann Intern Med* 1978;89:157-161.

271. Sullivan JL. Iron and the sex difference in heart disease risk. *Lancet* 1981; 1(8233):1293-1294.

272. Casale G, Bignamini M, de Nicola P. Does blood donation prolong life expectancy? *Vox Sang* 1983;45:398-399.

273. Skoog DA, West DM. Volumetric methods based on complex-formation reactions. In: *Fundamentals of Analytical Chemistry.* New York: Holt, Rinehart and Winston, Inc; 1969:338-600.

274. Peng CF, Kane JJ, Murphy ML, et al. Abnormal mitochondrial oxidative phosphorylation of ischemic myocardium reversed by calcium chelating agents. *J Mol Cell Cardiol* 1977;9:897-908.

275. Freeman AP, Giles RW, Berdoukas VA, et al. Early left ventricular dysfunction and chelation therapy in thalassemia major. *Ann Intern Med* 1983;99:450-454.

276. Blake DR, Hall ND, Bacon PA, et al. Effect of a specific iron chelating agent on animal models of inflammation. *Ann Rheum Dis* 1983;42:89-93.

277. Addonizo VP, Wetstein L, Fisher CA, et al. Medication of cardiac 00ischemia by thromboxanes released from human platelets. *Surgery* 1982;92:292.

278. Yagi K, Ohkawa H, Ohishi N, et al. Lesion of aortic intima caused by intravenous administration of linoleic acid hyperoxide. *J Appl Biochem* 1981;3:58-65.

279. Benditt EP. The origin of atherosclerosis. *Scientific American.* 1977 Feb:74-85.

280. McCullach KG. Revised concepts of atherogenesis. *Cleue Clin Q* 1976;43:247.

281. Mayron LW. Portals of entry—a review. *Ann Allerg* 1978;40:399-405.

282. Walker WA, Isselbacher KJ. Uptake and transport of macromolecules by the intestine: possible role in clinical disorders. *Gastroenterology* 1974;67:531-550.

283. Hemmings WA, Williams EW. Transport of large breakdown product of dietary protein through the gut wall. *Gut* 1978;19:715-723.

284. Rowe AH, Rowe AH Jr. *Food Allergy: Its Manifestations and Control and the Elimination Diets.* Springfield, Il: Charles C. Thomas; 1972.

285. Speer F, ed. *Allergy of the Nervous System.* Springfield, Il: Charles C. Thomas; 1970.

286. Dickey LD, ed. *Clinical Ecology.* Springfield, Il: Charles C. Thomas; 1976.

287. Randolph TG. *Human Ecology and Susceptibility to the Chemical Environment.* Springfield Il: Charles C. Thomas; 1962.

288. Crook WG. The coming revolution in medicine. *J Tenn Med Assn* 1983;76(3):145-149.

289. Iwata K. Toxins produced by *Candida albicans. Contr Microbiology Immunol* 1977;4:77-85.

290. Iwata K, Yamamota Y. *Glycoprotein Toxins Produced by Candida albicans.* Reprinted from Proceedings of the Fourth International Conference on the Mycoses. PAHO Scientific Publication 1977;356:246-257.

291. Iwata K. Fungal toxins as a parasitic factor responsible for the establishment of fungal infections. *Mycopathologia* 65:141-154.

292. Crook WG. *The Yeast Connection: A Medical Breakthrough.* Jackson, TN: Professional Books; 1983.

293. Truss CO. Tissue injury induced by *Candida albicans. Orthomolecular Psychiatry* 1978;7(1):17-37.

294. Truss CO. Restoration of immunologic competence to *Candida albicans*. *Orthomolecular Psychiatry* 1980;9(4):287-301.
295. Truss CO: The role of *Candida albicans* in human illness. *Orthomolecular Psychiatry* 1981;10(4):228-238.
296. Hatano S, Nishi Y, Usui T. Copper levels in plasma and erythrocytes in healthy Japanese children and adults. *Am J Clin Nutr* 1982;35:120-126.
297. Harrison W, Yarachek J, Benson C. The determination of trace elements in human hair by atomic absorption spectroscopy. *Clin Chem Acta* 1969;23(1):83-91.
298. Gori GB. Observed no-effect thresholds and the definition of less hazardous cigarettes. *Journal of Environmental Pathology and Toxicology* 1980;3:193-203.
299. Saltin B, Karlsson J. *Muscle Metabolism During Exercise*. New York: Plenum Publishing Co; 1971:395.

Periodic updated information will be posted on the Internet at:
www.drcranton.com
Dr. Cranton answers e-mail questions at:
drcranton@drcranton.com

A list of physicians who are members of the American College for Advancement in Medicine, a professional association of doctors who administer chelation therapy, may be obtained from:
American College for Advancement in Medicine
23121 Verdugo Drive, Suite 204
Laguna Hills, CA 92653
949-583-7666
Fax: 949-455-9679
www.acam.org
This is not an endorsement. Not all doctors are equally qualified.

Appendix A

CASE HISTORIES

I'll open this appendix of case histories with three narratives written by the patients themselves and addressed to a colleague of mine, Dr. Neil Scrimgeour, who practices EDTA chelation therapy in Bedford, Western Australia, near Perth. Dr. Scrimgeour forwarded these to me for inclusion here. I think they describe as well as I ever could the kind of transformation in health that a chelation doctor so frequently sees. It also occurs to me that I should really encourage some of my own patients to start writing. Personal testimonials of this kind have a quality of conviction that the most skillful third-person account can't equal.

From Dorothy A., a fifty-seven-year-old woman with severe coronary heart disease:
(Addressed to Professor T., her cardiology specialist after chelation)

I'm writing to bring you up to date with my progress.

When I last saw you I was on 12 heart and blood pressure tablets a day. I remember that you were slightly embarrassed that I was on every heart tablet you could give me.

After several months I was feeling very tired with a lot of pain from fluid in my lungs, tight chest all the time and the most stressful part, not enough air. I felt I would eventually suffocate. You had pointed out that the tablets were to give me a better quality of life, as you felt there was possible failure of the heart muscle and nothing could be done to help this. Since I felt no better on the tablets, I reduced them, still taking the blood pressure tablet, vitamins, and the hormone replacement.

Over the past four years we sold our home to avoid the stairs. I found I could not baby-sit or even pick up my grandchildren, even though I was only 57. I felt I couldn't cope. I remember one morning when I had showered and washed my hair thinking I had the fatigue of the dead.

A good friend came to see me and told me of Dr. Scrimgeour who did chelation therapy. An associate of his had had chelation with fantastic results.

After blood tests and a long consultation with Dr. Scrimgeour I commenced treatment.

After 10 treatments I felt no better and could have given up, but I saw many people experiencing positive improvements. These patients encouraged me to keep on.

After 14 treatments I realized my pain was not as bad and I could walk quicker—the load on my chest was less, and after two additional treatments:
• My coughing had stopped.
• I could walk around the shops without discomfort.
• I could clean the house quickly doing as much as I liked.

• I planted 60 roses digging the holes myself and carting the soil for them.

And then it happened.

I played 18 holes of golf pulling my own buggy.

Professor T., the joy was so great that I shed tears in the clubhouse shower after I came in. I had done 18 holes when I could not have done one before chelation.

My husband is ecstatic, as we had long ago given up hope that I would ever be able to walk like this again.

Please don't insult your or my intelligence by saying it was the hormone replacement patch or Vitamin E, as I had been on these for four years.

No side effects, either. I have spoken to many people. Some have not had as dramatic improvement as I have had, but they have all improved greatly, with no side effects.

There has never been a generation that has had to survive with so many metals, pesticides, toxins, and chemicals. We badly need the support of doctors like yourself that have not got tunnel vision.

Would I convince you [about chelation therapy] if I took you on the City to Surf fun run!!!

I would try anything. I feel wonderful. I honestly didn't know that living could be so great.

Dorothy A.

From Mr. Ronald D., who suffered with coronary heart disease since age thirty-one:

(The following is a report personally written by the patient for the records of Dr. Scrimgeour.)

Dear Dr. Scrimgeour:

In June 1983, at the age of 31, I first developed angina. I was admitted to the Hospital for cardiac catheterization that demonstrated 75 percent of the total coronary arteries were affected in one way or another, including a totally obstructed right coronary artery. A bypass surgeon at the Royal Perth Hospital performed a right coronary endarterectomy and reversed saphenous vein graft to the right coronary artery in October 1983.

I remained angina free until December 1984 [slightly more than one year] when I again experienced angina. It persisted until early August 1985 when I was rushed to Royal Perth Hospital following a prolonged episode of chest pain. There had been no myocardial infarction, but the frequency and severity of the angina showed new disease in the left coronary artery as well as the right—including a totally obstructed right coronary artery again. A second operation was performed in 1985 with saphenous vein graft to the lateral ventricular circumflex, a free internal mammary graft to the diagonal and the left internal mammary artery to the LAD [left anterior descending coronary artery], in other words a triple bypass.

In 1989 recurring angina got progressively worse. Another catheterization showed a totally obstructed right coronary artery and a severely obstructed left coronary artery.

I was referred to a cardio-thoracic surgeon at Royal Perth Hospital in September 1989 for bypass surgery using a gastroepiploic graft into the right and left coronary arteries. One end of a stomach artery was to be severed and looped up into the right coronary artery, then across to the left, but because of the time factor during the operation and the length of the gastroepiplocic artery, the left coronary artery was excluded. As a result, a few days following the operation, I still had chest pains. This operation showed gross disease to the right and left coronary arteries up to the crux of the heart.

I was discharged from hospital, slightly better but unfortunately still suffering from angina. By June 1992 repeated angina attacks and another angiogram indicated further artery disease, but further bypass surgery was decided against.

Till now the angina was caused by exertion or extreme conditions and was easily controlled by nitroglycerine or rest. But my condition deteriorated and eventually I required 25 nitroglycerine tablets daily.

In 1994 I was back in hospital after a severe angina attack. That night while hospitalized I experienced another severe episode and had to be given a streptokinase injection to dissolve the clot. An EKG indicated that I had had an inferior infarct. Further testing showed a return to normal. I was lucky!

In 1995 and 1996 despite maximum medical therapy, I was experiencing between 45 and 50 severe and frightening episodes of angina a day, consuming approximately 50 nitroglycerine tablets daily and several calcium channel blocker capsules. These chest pains were horrendous and lengthy. I was unable to walk, move around or go outside without having angina pains. I could only manage a few feet at a time. Lying down to rest or sleep was a problem and a nitro patch had to be used each night. My severe limitations caused weight gain.

In May of 1997 a further catheterization revealed major artery problems, but re-do surgery was considered to carry significant risk.

At this time, my parents informed Dr. Neil Scrimgeour of my review and he suggested to them that I should consider chelation therapy. I started Dr.

Scrimgeour's chelation therapy and after only three chelation sessions I was relieved of severe and extreme chest
pains. My intake of nitroglycerine declined from 50
tablets daily to 18. After nine sessions I was taking six
tablets daily and managed to start light exercise. I continued with chelation therapy and am now able to get
through the day with a minimal amount of nitroglycerine
and, at times, none. Unbelievable! I am now able to go for
daily walks and manage a 60-minute exercise workout
each morning, losing a remarkable 16 kilos [35 pounds].

Chelation therapy has completely changed my life,
and if I had known about it, would have saved me from
past surgeries. I am totally convinced of that. I thought
when I began that if I lost only a small percentage of
chest pain that would be a big relief. The results far
exceeded my expectations, along with those of certain
other medical personnel who were absolutely amazed
with the results.

Ronald D.

*From Reginald S., a sixty-year-old retired soldier with severe
coronary heart disease:*

Dear Dr. Scrimgeour:

I feel I should write to you and give you a detailed
account of myself before, during, and after chelation
treatment. If you think it will help your new patients,
please pass this letter around.

I was able to retire at 60 because I was a returned
soldier. I visited my local doctor yearly and was pronounced healthy.

I went up north every winter to where the snapper
up to 14 pounds came in on high tide. It was not an
easy task landing the big ones. I wondered why I was

getting chest pains but passed it off as indigestion. One day my mate and I were walking up the beach to another fishing spot, and I had to sit for a rest. I explained to my friend what was going on, and he soon convinced me I was getting angina.

When we got back a doctor arranged for me to enter the hospital. My condition was so bad I was not allowed to walk anywhere.

I was tested and told a five-way bypass operation would put me right. Then I was sent home to wait my turn and put on 16 tablets of heart medicines per day. Two heart attacks later I was admitted to the Royal Perth Hospital and had a six-hour operation. I was discharged after three weeks.

Two days after discharge I was rushed back to hospital with another heart attack. Soon after I went home to wait for another operation. If it was like the first one, I felt sure it would kill me.

I could only walk to the front gate and back to a chair or bed. My family and friends called me a walking corpse. My whole body felt cold, I felt that death was not far away. A few weeks later my wife went to the chiropractor and came back all excited saying, "I have some good news for you—there is a treatment called chelation which could save you from having another bypass operation."

I could not get to you quick enough, Dr. Scrimgeour. You gave me tests and remarked that I was the coldest person you had ever treated and by the time I had 20 treatments I would be much warmer. This proved to be correct.

My sister drove me to the clinic three days a week for chelation—walking to catch buses and trains was impossible. After six treatments she could not get over how the color of my skin was getting back to normal. After nine treatments my sister did not have to drive me. I was able to walk two blocks and catch a bus.

After 20 treatments I was starting to feel my old self again—not one tablet did I have to take. My family and friends could not get over how well I looked. I slowly took up golf again, looked after my veggie garden, and was able to do my home duties.

You advised me to return for six chelation treatments a year to keep me in good order.

All was going well, but not long after that I got a message to say there was no more chelation. The government had stopped the drug EDTA used in chelation from being allowed into Australia.

Well! I reckon that was the lowest thing the powers that be could do. My life in my mother's womb was short, so what with no chelation my life in the world was meant to be short also. I made up my mind to arrange and pay my funeral expenses. I went along to a funeral director, picked out my coffin and asked how much would it cost. "$4,500," he said. "Bugger you, mate," I said, "I'll make myself a coffin." (As it's very rare to see, I have enclosed photos.)

Time went by—in fact, I had not finished making my coffin, when I received news that chelation was going to start again. That was wonderful news!

It's eleven years now since I started chelation, and I have not needed my coffin yet. In all that time, I have not taken one heart tablet or had any other treatment for my heart problem.

I hope this letter will help people to make up their mind to have chelation before [or after] a bypass.

Yours sincerely,
Reginald S.

That ends the reports from our friends down under.

Jack F., now sixty-six years old, was successfully chelated under my care after three bypass operations plus angioplasty had failed. Before chelation, he had been referred for heart transplant.

Jack has suffered with coronary heart disease for twenty-five years, since the age of forty-one. In 1975 he first experienced onset of anginal chest pains, leading to angiograms and a two-way bypass operation. Following that first operation, Jack's symptoms improved for seven years and he was able to return to work.

In 1982, however, angina returned, and he underwent bypass surgery for a second time—unsuccessfully. Symptoms were partially relieved for only a brief time, and by 1987 he was totally disabled. In 1988 a third bypass operation was attempted, this time using the internal mammary artery instead of a vein graft—since all of his veins had been used during the two previous bypass operations. The third bypass surgery was totally unsuccessful—Jack received no relief at all following that operation. He continued to live with severe symptoms, taking many different types of cardiac medicines, totally disabled and unable to work.

In 1989 balloon angioplasty was attempted, with only minor relief lasting for a few months. In 1993 Jack had a full-blown heart attack—a myocardial infarction. No further surgery or angioplasties were attempted.

By 1994 he was taking twenty nitroglycerine tablets, plus forty tablets per day of a dozen other types of heart medicines. He was in congestive heart failure and traditional medicine had nothing further to offer. In addition to heart disease, he also had symptoms of claudication (leg pain on walking) caused by arterial blockages.

As a last resort, Jack was placed on the waiting list for a heart transplant at the University of Washington Medical

Center in Seattle. That's when I first met him. He had heard about chelation therapy and came to me for treatment.

Between 1994 and the end of 2000, Jack has completed seventy-two chelations with remarkable benefit. He's still not totally well but uses nitroglycerine only occasionally. His heart failure has been reversed, and the quality of his life is better than it was a decade ago. Within a few months of starting chelation therapy, his condition improved enough that he was removed from the waiting list for a heart transplant. His cardiologist, a medical school professor, seemed very surprised at Jack's unexpected improvement and advised him to continue with chelation therapy.

In Jack's own words, "Chelation helped me a hell of a lot." He's now able to walk without angina or leg pain, and he recently traveled across the country to visit relatives. He plans to drive to Alaska next summer, regularly does work in his shop, and goes fishing—things he would not have been able to do before chelation. Jack returns periodically for follow-up treatments and continues to maintain his improvement.

Bob H. is a fifty-year-old clinical psychologist who suffered from severe scleroderma.

When he first came to me, Bob H. was a professional who regularly worked with patients dealing with life-threatening illnesses. He had always been in excellent health, exercised regularly, and followed a healthy diet.

When Bob first experienced unusual symptoms, he did not take them seriously. After all, he could not possibly be sick—but the symptoms continued and even worsened. By the time of his first medical consultation, he had progressed to diffuse muscle stiffness, thickening and hardening of the skin such that he was unable to fully close or open his hands, fatigue, swelling of hands and feet, sensitivity to pressure when touched, and difficulty swallowing from sclerosis of the internal digestive tract.

The event that finally spurred him to see a doctor occurred in the staff cafeteria at his local hospital. Bob was eating lunch with a group of doctors on staff at the hospital, including two surgeons, when he began to have an esophageal spasm. The muscles in his throat tightened and he could not swallow. At first he was fearful he couldn't breathe, but quickly realized that he could. He then became fearful that he would pass out and wake up to find the surgeons had performed a tracheotomy.

That was it. Bob sought medial help. But after seeing four different physicians, he still had no answers. All agreed that something was going on, but no one had a firm diagnosis. Then he was off to a major medical center for a long series of medical tests and examinations. Finally the specialist told him that he had systemic sclerosis (scleroderma, diffuse type). This is generally recognized as the deadliest form and rapidly progressive.

Undaunted by the diagnosis, Bob set out to get the best medical treatment available. After many more doctor visits and a lot of time in the medical library, he became discouraged. According to the published research, he had a disease with no known cause and no known cure. All that modern medical science had to offer was penicillamine and prednisone (an oral form of cortisone). Convinced that penicillamine had little effect on scleroderma (with severe side effects), and prednisone would be the beginning of the end, Bob began to investigate alternatives. That's how he came to my clinic.

After reading all the information he could find on chelation, Bob was convinced that I had more hope to offer than conventional treatments. By the time I first saw him, his symptoms had become much worse. The skin on his face, hands, arms, and legs was very tight and shiny. He described the way his skin felt as though he had "walked bare-chested through spider webs," and when he moved the webs were pulling. He was very stiff and could not walk one hundred yards without becoming extremely fatigued and out of breath.

I did not have to convince Bob to try chelation; he was sold on the treatment and ready to start. At first he noticed no change at all, but he nevertheless wanted to continue. After thirty sessions he reported the spider-web feeling was decreasing. He then experienced a gradual clearing of symptoms over a two-year period. He was able to return to a full schedule of work, and all of his other activities were resumed. Bob has now received approximately 130 chelation treatments over the past eight years. He plans to continue getting twenty or so chelations per year to maintain his benefit.

With the exception of some residual inability to completely open his hands, Bob has remained generally symptom-free for seven years. In his own words, "If it had not been for chelation therapy and the nutritional supplement program you recommended, I would not be here today." Bob has become a strong advocate for the use of chelation in scleroderma.

Whether they be brief or long, case histories almost always command interest. Many were interspersed through the text of this book, but it was not possible to include as many as I would have liked. So I've added a few more here in abbreviated form to satisfy the curious. I would like to thank the many patients who so generously allowed us to use their case histories.

A sixty-eight-year-old man with severe heart problems:
Charles S., a minister, had his first heart problem in 1989. Over the course of the next eight years, he underwent a dozen angiograms with multiple balloon angioplasties and two bypass surgery operations. In 1997, after he was diagnosed with yet more severe arterial blockages and recommended for more surgery, he opted instead for chelation therapy. His symptoms quickly improved. He no longer suffered shortness of breath, edema, angina, and fatigue. His symptoms all regressed and now, three years after completing fifty-five chelations, he can mow his own lawn and play vigorously

with his grandchildren—simple activities that were beyond his power before chelation. And he has avoided further surgery.

A sixty-five-year-old man with multiple arterial blockages:
George W. is a retired seaman who had his first carotid artery operation at the age of sixty-one. A year later, after suddenly losing vision in one eye, he was told that he was primed for a stroke and was subjected to another carotid endarterectomy. He survived the second operation with difficulty—as the surgeons sometimes say, he had a "stormy course." By then, however, he had also developed intermittent claudication—pain on walking from arterial blockage in his legs. He was unable to walk more than ten feet without pain. He began chelation therapy in March 1998. Several months and thirty chelation treatments later, he experienced a return of overall vitality. His walking distance increased until today he can briskly walk for three miles or more without stopping. He enjoys an active lifestyle and is entirely free of his former symptoms.

A sixty-five-year-old woman with heart disease:
Harriet G. is a retired office worker who suffered from disabling coronary heart disease before chelation. She suffered anginal chest pains during minor physical exertion and had high blood pressure. A coronary arteriogram showed serious blockage. She declined bypass surgery, and in March 1999 she began a series of thirty-three chelations. She is now totally free of angina and is able to lead an active life. Her blood pressure is lower, and she no longer takes nitroglycerin. She's delighted to be able to care for her family and travel again.

A fifty-seven-year-old man with heart disease:
Arthur D., a sales representative, suffered with occlusive coronary artery disease, carotid blockage in the arteries to his

brain, high blood pressure, and disabling chest pain on exertion. In the summer of 1997, he had a severe heart attack (myocardial infarction). Cardiac catheterization and arteriogram were done, after which his doctors told him that three of the major arteries to his heart were 90 to 95 percent blocked. They proposed immediate bypass surgery. Arthur was uneasy with that recommendation, however, and having heard about chelation from a nurse practitioner, decided to try it. He has since taken forty-five chelation treatments and is now completely symptom-free. He walks several miles every day. By his own account, he can now do almost anything he wants and no longer experiences angina.

A sixty-year-old woman with coronary heart disease:
Charlotte A. suffered exertional chest pain and was taking many medications for coronary heart disease. After walking only one block, she consistently had onset of chest pain severe enough to cause her to stop. She frequently needed to take nitroglycerin tablets. Charlotte began chelation therapy early in 1999 and now walks four miles a day without angina and with no further need to take nitroglycerine.

A seventy-two-year-old man with coronary heart disease:
Harold A., the retired general manager of a power company, developed serious cardiovascular problems in the early 1980s, and in 1981 had a five-vessel bypass operation after suffering two myocardial infarctions. By 1987 his heart problems were again serious. He experienced shortness of breath and disabling chest pain after minimal exertion. Wary of further surgery, his wife suggested chelation. In 1988 Harold began a series of chelation treatments. During the years since, he has received a total of 125 chelations and no longer suffers angina. Now, thirteen years later, he continues to lead a healthy vigorous life without limitation.

A sixty-seven-year-old man with heart disease:

Paul C. had a long history of heart disease that went into remission after chelation. Paul had his first heart attack in 1977 at the age of forty-four. He improved his diet and followed his doctor's advice, but in 1979 he underwent his first bypass operation. His second bypass came in 1988. In 1992 a coronary cardiac catheterization and arteriogram showed that his arteries were closing down again. Further surgery seemed necessary. Exhausted and in pain, Paul became totally disabled and forced to quit work. It was at that point that he first read an earlier edition of this book, and in January of 1993 he began chelation, starting with 30 treatments. His symptoms soon improved and he returned to work. Now, seven years later, he has completed 120 chelation treatments and leads a full life with no further heart symptoms.

A seventy-year-old man with severe heart disease:

Jim C. had a long history of coronary heart disease that began in the 1970s. In 1980 he had a quadruple bypass. In 1990 he suffered a severe heart attack. His condition became progressively worse, and by 1999 he was almost too disabled to continue with his business as a locksmith. Jim suffered from severe shortness of breath and anginal chest pains on exertion, continuous fatigue, and was failing at a rapid rate. Fearful of undergoing another bypass operation, he began chelation. After thirty treatments, not only was his electrocardiogram improved, but he could actually run up a flight of stairs without chest pain or shortness of breath. His symptoms have completely resolved, and, in his own words, "I feel better than I have in years."

I'll now add three case histories of patients who benefited from hyperbaric oxygen therapy (HBO), which was discussed only briefly in chapter 13. I have found it to be an excellent adjunct to chelation therapy in many cases. For the last three case histories, I am indebted to Pavel I. Yutsis, M.D., of Brooklyn, New York, who routinely uses this therapy in his clinic.

A forty-eight-year-old man with critical diabetic gangrene and threatened amputation:

Reber T. first came to my clinic in 1982 after being told by two different surgeons independently that he must have his left leg amputated or he would die. Reber suffered with diabetic gangrene of his left foot, and infection was spreading up his leg to the knee. Two toes were black. His leg below the knee was red and swollen with cellulitis. He was in constant severe pain—but he was stubborn and resisted amputation.

In Reber's case, we combined hyperbaric oxygen treatments (HBO) with chelation therapy. HBO helped the antibiotics combat infection and kept oxygen-starved tissues in his foot alive, giving chelation therapy time to bring in new blood flow. The infection gradually receded, and the foot became pink and warm as new blood flow was established. The two black toes were eventually lost, but the leg and foot were saved. Now, eighteen years later, Reber continues to walk and work on his own two legs.

A sixty-four-year-old man with sudden blindness in an eye from carotid plaque embolism:

Herbert W. quite suddenly lost vision in his right eye. After conferring with his family physician, he was referred to an ophthalmologist for evaluation. He also got a second opinion from another ophthalmologist. Both eye specialists told him the same thing—he had suffered arterial blockage to his eye, and the eye was permanently blind as a result. They had no treatment to offer. Within three days he was at my clinic to begin hyperbaric oxygen therapy. After his second HBO treatment he was able to make out some detail. With further HBO, vision in his bad eye subsequently improved to 20/40 and remains so today, seven years later.

A man with brain damage from cerebral hemorrhage treated with HBO:

Mr. S. B. is a sixty-year-old man who suffered a major stroke from cerebral hemorrhage five months before entering treatment. He was in a coma for one month, and a blood clot was surgically removed from the back of his brain. He was left with slurred speech, impaired memory, poor balance and coordination, and he needed a cane to walk.

He completed a course of forty-five HBO treatments, and all of his functions returned to normal, with no residual symptoms.

Another man with brain damage from cerebral hemorrhage treated with HBO:

Mr. J. G. is a thirty-six-year-old police officer who suffered a cerebral hemorrhage more than four years prior to receiving HBO. A large collection of blood had been surgically removed from the left side of his brain. Following this he suffered with residuals of poor memory, difficulty speaking and finding words, severe right-sided paralysis, and he could walk only with a cane. This patient received a total of 125 HBO treatments, after which his speech became understandable, his memory was much improved, and the right-sided paralysis improved significantly. Although not totally well, he was once again able to walk without a cane and could play baseball with his son.

A man with cerebral palsy, complicated by cerebral hemorrhage, treated with HBO:

Mr. R. Z. is a fifty-eight-year-old pharmacist who suffered with life-long cerebral palsy, manifested by partial paralysis and spasticity of the right side of his body. Three years before receiving HBO, he experienced a cerebral hemorrhage, leading to further paralysis affecting his left side, requiring a wheelchair. He could no longer swallow and was fed through a tube in his stomach. He had great difficulty pronouncing words because of paralysis of speech muscles.

After 220 HBO treatments, Mr. R. Z. could once more swallow and eat normally. The tube was removed from his stomach. His mental clarity improved. He discontinued using his wheelchair and could walk with some assistance. His ability to talk improved also, spasticity decreased, and he could once again care for himself without nursing assistance.

Appendix B

NUTRIENT-DEFICIENT FOODS— A MANMADE PROBLEM

This chapter will illustrate two instances of what's happening to our modern food supply. Remember that much of what will be true for these two examples will also be true for many other foods and nutrients. For most people, the diets of our pets and livestock are more nutrient-rich than our own diets. One reason this is true is the "bottom line" factor. If livestock are not relatively healthy and do not grow adequately, the producer makes little or no profit. So, generally speaking, healthy livestock equals healthy profits. It's an irony of life that we do not translate that equation into "healthy people equal a healthy and productive society."

Bread, a Degraded, Modern-Day Food

Modern technology has transformed bread, once the staff of life, into a mere broken reed, contributing to widespread vitamin and mineral deficiencies. This occurs even in those wealthy, industrialized nations where food is plentiful. Bread is described in this appendix as just one example of similar processes that degrade our entire food supply on its path from farm to consumer.

In order to obtain the advantages of high-tech food processing, mass-production, mass-marketing, long shelf-life, uniformity of final product, even coloration, and soft texture, the baking industry has become a source of widespread nutritional deficiencies. The food-manufacturing industry deceptively markets its products as more convenient and tastier versions of the bread that grandmother once made in her kitchen. That's far from the truth!

Today's mass-produced foods have largely been depleted of nutrients and are highly chemicalized with additives. The processed foods of today are not just more sophisticated and more convenient versions of the foods eaten by our ancestors. A wide spectrum of essential nutrients has been removed in the manufacturing process. The basic molecular structure of what remains has also been degraded to become nutritionally inferior.

Until recently, grains were ground between large stones to make flour. Everything in the original grain remained in the finished product, including the germ, the fiber, the starch, and a wide variety of vitamins and minerals. The final product contained all of the naturally occurring and essential vitamins, minerals, and micronutrients.

In the absence of refrigeration, stone-ground flour spoils quickly. After wheat has been ground, the wheat-germ oil becomes rancid at about the same rate that milk becomes sour. Whole-wheat flour and bread should therefore be stored in a cool place, preferably in a refrigerator.

Hippocrates, a famous physician in ancient Greece, once recommended high-fiber stone-ground flour for its beneficial effects on the digestive tract. Today, three-fourths of that dietary fiber is removed from commercial flour when milled. As a result, constipation is a common ailment in this modern culture.

During the industrial revolution of the nineteenth century, assembly-line techniques for mass producing flour and bread were first developed. Stones for grinding were not fast enough for mass production. High-speed, steel roller mills were invented to produce flour very rapidly. The milling company thus earned higher profits.

High-speed mills do not grind the germ and the bran properly. They are ejected. Those constituents of the original grain, actually the most nutritious portion, are thus removed and sold as "byproducts" in animal feed. As a result, animals are better nourished than people. It's been cynically observed that the greatest profit comes from healthy animals and sick people.

When baked, the interior of bread does not get much hotter than 170 degrees, which is not greatly destructive of vitamins. High-speed mills have changed that—they run very hot, at 400 degrees Fahrenheit, just under the temperature that will scorch and discolor the flour. That high heat destroys many vitamins.

Since the late nineteenth century, white bread, biscuits, and cakes made from white flour and sugar have become mainstays in the diets of industrialized nations. The result is a diet much less nutritious than customary in former times. This deficiency has caused new diseases of affluence. Tooth decay, once rare, is now epidemic. The incidence of tooth decay correlates exactly throughout the world with industrialization and the use of refined foods—especially white flour and sugar.

For the most part, bread is now manufactured in large factories capable of producing up to a quarter million loaves per day. This mass-produced bread is soft, gooey, devitalized, and nutritionally deficient—laced with chemical additives. Public taste is accustomed to such bread. People have forgotten how real old-fashioned bread tastes. Chemical preservatives allow bread to be shipped long distances and to remain unrefrigerated on the shelf for many days without spoiling, again resulting in less product loss and higher profits.

To make bread a brighter white, at the expense of consumer health, flour is treated with chemical bleach, similar to Clorox. The bleaching process leaves residues of toxic chlorinated hydrocarbons and dioxins. Methionine, an essential amino acid, reacts with bleaching chemicals to form methionine sulfoxine, a toxic residue that causes nervousness and seizures in animals.

The bleaching process further destroys many vitamins (those not already destroyed by the high heat of milling). Bleaching agents have therefore been banned from use in bread making in Germany since 1958. In the United States, however, no such ban exists, and bleached white bread continues to be the mainstay. Most white flour now used to make bread, rolls, cakes, pastries, spaghetti, noodles, pasta, and breakfast cereals has been bleached.

Grain millers in the nineteenth century soon discovered that highly refined flour would keep much longer without spoiling, even before the days of chemical preservatives and refrigeration. It's now clear that refined flour is so depleted of essential vitamins, minerals, and other micronutrients that it will not support life. Even the insects and rodents can't live on it! Can humans be expected to fare any better?

Experiments were reported in a major British medical journal, *The Lancet*, showing that dogs fed exclusively on white bread died of malnutrition within two months. Dogs similarly fed only bread made with stone-ground, whole-wheat flour lived indefinitely in good health.

Chemicals are added to supermarket breads in large quantities, despite increasing reports that similar chemicals previously thought to be safe are potential causes of cancer. More than thirty different chemicals have been approved by the Food and Drug Administration (FDA) for addition to bread, including ethylated mono- and triglycerides, potassium bromate, potassium iodide, calcium proprionate, benzoyl peroxide, tricalcium phosphate, calcium sulfate, ammonium chloride, and magnesium carbonate. These are added to bread to extend shelf life and improve uniformity and texture, despite the fact that little is known about their long-term cumulative toxicity when taken together. If you don't already read labels, you'll be shocked if you do.

When grain is made into refined white flour, more than thirty essential nutrients are largely removed. Only four of those nutrients are added back in a process called "enrichment." Using this same logic, if a person were to be robbed of thirty dollars and the thief then returned four dollars to his victim for cab fare home, that person should be considered "enriched" by four dollars, not robbed of twenty-six dollars. How would you feel in that situation?

You should feel the same about "enriched" white flour and bread. Only vitamins B1, B2, B3, and iron are replaced. Nutrients that are removed and not returned include the following: 44 percent of the vitamin E, 52 percent of the pantothenic acid, 65 percent of the folic acid, 76 percent of the biotin, 84 percent of the vitamin B6, and half or more of 20 minerals and trace elements, including magnesium, calcium, zinc, chromium, manganese, selenium, vanadium, and copper. If consumers would educate themselves in the principles of good nutrition and act on that educated preference at the checkout counter, the food industry would be forced to respond with more nutritious products.

Iron, the single mineral added back to enriched white flour, is already present in excess amounts in the bodies of many older people. Excessive iron contributes widely to atherosclerosis, heart attacks, strokes, senility, arthritis, cancer, and other age-related diseases. It's quite possible that enrichment of flour with iron has been poisoning the public for decades. Although iron is occasionally needed to treat deficiencies, avoidance of unneeded iron supplementation is reason enough by itself to avoid so-called "enriched" flour products.

Deceptive marketing practices are widespread. Much of the bread now marketed as "whole-wheat bread" is made with the same old refined white flour, with a little brown coloring added. That coloring is usually burnt sugar, listed on the label as caramel. One manufacturer even added sawdust to replace the lost bran, calling it cellulose on the label and advertising it as "high-fiber" bread. It is perfectly legal to describe inferior flour as "whole wheat" on the label, even when most of the bran and germ have been removed in high-speed roller mills.

It is much slower and more expensive to mass-produce bread made with 100 percent stone-ground whole-wheat flour. Manufacturers go to great lengths to mislead the public by making inferior products appear to be of higher quality. Without chemical preservatives, bread spoils rapidly. It quickly becomes stale, hard, and moldy. To transport nutritious whole-grain, unrefined bread over long distances is not practical. It would require refrigerator trucks for delivery and refrigerator storage in supermarkets. Even under refrigeration, whole-wheat bread would spoil faster than chemicalized bread. That loss would add greatly to expense. Profits would be smaller. Production of truly nutritious bread therefore falls to small local bakeries, which sell direct or deliver daily to nearby stores.

Low fiber in a diet of refined flour is one cause of bowel cancer. Without bran, transit time through the digestive tract is greatly lengthened. Constipation results, causing hemorrhoids, diverticulitis, and increased risk of colon and rectal cancer.

What's the solution? Ideally, one should buy wheat in sacks, grind the grain at home, and quickly bake it into bread. An alternative would be to buy stone-ground whole-wheat flour at a natural food store, ground at the time of purchase on the premises, refrigerated at home, and used quickly. Stone-ground flour will keep fairly well for several months frozen and for a week or more refrigerated.

Unfortunately, most of us no longer have time in our schedules for baking at home and must rely on store-bought products. To determine which bread is best, read the label thoroughly and choose a product that has the brown coloring of natural flour, without any coloring agents added. Choose a product with a minimum of chemicals listed on the label. Whole-grain bread does not rise as much and therefore contains more wheat and less air. A good loaf will be proportionately heavier, firmer to squeeze, and chewier. It will also cost more, but you get more for your money, and the flavor will be much better.

Slow-speed steel-hammer mills are often used instead of stones. That type of flour may be listed on the label as "stone-ground." It's fully equivalent to stone-ground flour and is equally nutritious. Any process that renders the entire grain into usable flour without exposing it to high heat is acceptable.

If a loaf made with 100 percent stone-ground flour cannot be found, choose one with unbleached flour. "Gluten flour" is just another name for partially refined flour but is better than bleached white flour. So-called "unbleached whole-wheat flour," which is processed on high-speed roller mills and used in most so-called whole-wheat bread, is missing much of the vitamins, bran, and germ.

If bread is made only with 100 percent stone-ground whole grains, that fact will be stated on the label. If the label does not contain that statement, then you must assume otherwise. Many bakers add refined or so-called gluten flour to produce a lighter and more uniform product. It's difficult to make the loaves come out looking alike with stone-ground flour.

Unbleached flour is better than bleached but is still inferior unless it is 100 percent stone-ground. Bakeries seldom state the exact percentage of whole-grain relative to refined or unbleached flour. In those instances, it's usually safe to assume that very little stone-ground, whole-grain flour is used.

A search through grocery stores and supermarkets today will rarely reveal a loaf of bread that meets all the criteria for good nutrition. However, many small bakeries exist that produce superior products for local sale, either direct or in natural food stores. Read the labels carefully. Just because a product is sold in a health food store does not ensure that it is of high quality.

Look for a loaf that states "only 100 percent stone-ground whole-wheat flour" on the label. Refrigerate it. Expect it to be heavier and chewier. Squeeze it. If your fingers go in easily and the bread springs back, it is not a fully nutritious loaf. If you don't eat it within a few days, freeze it until needed. Expect to pay more. Whole-grain bread does not rise as much and contains more wheat than the same size loaf of refined bread. You are paying for more grain, more time in production, and less air. You will be nourished and get much more for your money.

Sugar, Chromium, Minerals, and Trace Elements

Metallic elements in the diet (commonly referred to as minerals) are equally important to human health as are vitamins—perhaps even more so. Minerals required in tiny amounts are more correctly called trace elements. Except for calcium in bones and iron in red blood cells, the importance of a wide variety of other dietary minerals in human health and nutrition is commonly overlooked. For the same reasons that vitamin deficiencies are so prevalent, even in wealthy industrial nations, deficiencies of essential nutritional minerals are also common. Very few people obtain optimal amounts of all micronutrients.

Laboratory tests show that, when compared to our ancestors, almost all Americans receive below optimal levels of nutritional minerals and trace elements (other than iron, which is often present in excess). Mineral deficiencies contribute to many diseases and premature aging. Toxic elements become much more toxic with coexisting deficiencies of the nutritional elements—a factor that greatly compounds the problem. Industrial pollution of the environment and food supply contributes greatly to the accumulation of toxic metals such as lead, mercury, and cadmium in the body.

I'll now focus on just one essential trace element, chromium, as an example of what happens to many other nutrients. The chromium story could be written in a slightly different way and applied to a dozen or more other essential nutritional minerals.

The Standard American Diet, disparagingly referred to as the "SAD" diet, contains a large percentage of highly refined and processed foods. Such manufactured foods have little in common with foods in their freshly harvested state, as they first come from the farm. More than half the vitamin and mineral content is lost in the refining process. In some cases, the final product contains less than 20 percent of the original amount.

A high-fat diet also contributes to deficiencies. In the United States, on the average, more than 45 percent of all calories are consumed as fat. Fat contains far fewer minerals than other foods.

Agricultural soils have been used over and over, year after year, for decades or even centuries without replacing essential nutritional minerals and trace elements. Chemical fertilizers and airborne pollutants further reduce incorporation into plants of the depleted trace minerals remaining in the soil. As a result, plants grown on depleted soils and animals grazed on those soils are already deficient at harvest, long before the food-processing industry can remove more of the little that did manage to be absorbed. Increasing sulfur fallout from the sky in rain water (acid rain from fossil fuels) competes with selenium for plant uptake. As selenium is depleted by successive harvests, the little remaining cannot compete with increasing concentrations of sulfur—compounding the problem of selenium deficiency.

Those many factors are cumulative and result in severely depleted foods as they reach the consumer in the grocery store.

On the average, Americans derive more than 65 percent of their calories from sugar, white flour, and fat, which leaves only one-third of the food still in relatively complete form, retaining its unrefined amounts of the many micronutrients essential for health.

Chromium: An Example of Trace Element Depletion

Plants don't need chromium. People do. Chromium is essential only for the humans and animals who eat the plants. Farmers therefore have no incentive to replenish chromium and similar trace elements back into the soil. Their harvest and profits are not improved by incurring this added expense. Each year's harvest has removed another portion of what little chromium remains, until at the present time agricultural soils have very little remaining. Unless crops are grown on organic, composted soil they will contain very little chromium and similar trace elements essential to human health.

Fifty thousand enzymes in the human body each require the presence of a specific metallic element (mineral or trace element) to function. Enzymes are the sparkplugs of cellular metabolism, without which cellular metabolism is not possible. Essential nutritional trace elements include manganese, copper, zinc, cobalt, molybdenum, chromium, selenium, vanadium, boron, and others. Even arsenic, a known toxin, is also essential for human health in ultratrace amounts.

Nutritional minerals also include calcium, magnesium, sodium, potassium, and iron, but in much larger amounts. (Iron in the so-called "fortification" process and sodium as salt are added to so many foods that deficiencies are rare and excesses are common.)

Normal concentrations of many trace elements in body tissues are so minuscule that they are measured in parts per billion. Even in such tiny concentrations, however, those elements perform vital functions and are necessary for life and health. On the other hand, those same elements are toxic and will cause illness if taken in excess. Just the right amounts are needed in the just right place and in just the right form, which is usually in an organic complex.

Deficiencies of trace elements have been implicated in many degenerative diseases associated with aging, including diabetes, heart attacks, arthritis, strokes, senility, and even cancer. The technology used to measure such tiny concentrations has only recently been developed. Very sophisticated instruments, using atomic absorption spectroscopy and inductively coupled plasma emission spectroscopy, are fairly recent developments. Nutritional research using those instruments has only been published in recent years and is not routinely taught in medical schools. Laboratory tests that allow doctors to test their patients for trace element deficiencies are new, and interpretation remains uncertain.

Virtually all Americans are deficient in chromium. Necessary chromium intake is so small, millionths of an ounce per day, that it would take four hundred years to safely consume one ounce. Even so, without chromium there would be no human life. On the other hand, too much is toxic. Chromium cannot be forced in by using large doses as can safely be done with water-soluble B-complex vitamins. It takes years to replenish a deficiency. A healthy person requires just the right amount, in just the right chemical form, with just the right valence (electrical charge), as is found in natural, unrefined, organically grown foods.

When laboratory animals are fed a diet low in chromium, they soon develop high blood sugar or diabetes. They also develop high blood cholesterol and hardening of the arteries (plaque formation called atherosclerosis), which is the leading cause of death in the industrialized world. Atherosclerosis causes heart attacks and strokes. When animals are fed a diet abundant in chromium (not too much, just the right amount), they live longer than animals on the usual laboratory diet, and they develop no evidence of diabetes or hardening of the arteries.

Humans who live in highly industrialized nations and who consume diets with a high percentage of fat and refined foods, white flour, and sugar are quite deficient in chromium. Chromium measurements in body tissues of modern-day inhabitants of the United States are only one-fifth the levels measured in people who live in less developed parts of the world. Populations who eat a diet of natural unrefined food, wild game, and food from virgin soils have chromium levels five times higher. And they experience very few deaths from heart attacks, strokes, or diabetes.

Chromium in arteries of victims of accidental deaths is significantly higher than that of people who die from heart attacks or strokes. Often, no chromium at all can be detected at autopsy in patients who die from atherosclerosis or diabetes.

Chromium has been measured in mummified remains of people who died centuries ago, before food refining and processing became widespread. Tissues of present-day Americans contain less than one-sixth or approximately 15 percent of the chromium found in tissues of human ancestors who lived long ago, before food was tampered with by technology and before agricultural soils became depleted.

Chromium is absolutely essential in order to utilize sugar, starch, and other carbohydrates. Without chromium, insulin cannot function and carbohydrates cannot enter cells as fuel for energy. When sugar and starches are eaten, chromium is released into the bloodstream from the liver to allow insulin from the pancreas to function. During that time, some of that chromium in the blood is lost in the urine. If refined sugar and white flour are eaten, not enough chromium is present to replace what is lost. Whole foods grown on fertile soils contain enough chromium to replace the amount lost in urine following a meal of carbohydrates. A positive balance is maintained, and body stores are preserved. A diet of refined and commercially processed foods, with sugar-laden soft drinks, causes a slow, progressive loss of chromium throughout life, contributing to diabetes, hypoglycemia (low blood sugar), atherosclerosis, and premature death.

At ten years of age the chromium concentration in tissues of an average American child is less than 20 percent of what was present at birth. That is largely because refined sugar consumption has increased from less than ten pounds per person per year two hundred years ago to an average of 120 pounds per person per year in 1994. An average American now eats his or her weight in sugar each year. As a result, chromium is progressively lost throughout life, blood cholesterol increases, and, without chromium, the pancreas is forced to produce increasing amounts of insulin to allow blood sugar (glucose) into cells to be used as fuel. Even if diabetes does not develop, increased insulin has been incriminated as a cause of obesity, high blood pressure, and heart disease.

Insulin levels come down and the symptoms of diabetes decrease when patients improve their diets to include more chromium-rich foods. What foods are high in chromium? Well, it seems that Mother Nature, in her wisdom, saw fit to include adequate chromium in all naturally occurring sugars, starches, and whole grains when they are grown on fertile, composted soil. It's only when we remove chromium by food processing and allow our agricultural soils to become depleted that those problems occur.

Sugar cane has four times the chromium per calorie of refined white sugar. During sugar refining, that chromium ends up in molasses, which is not consumed in any quantity by humans. It's more often fed to animals as a byproduct. Honey has five times more chromium per calorie than sugar. White flour has only one-fourth or 25 percent the chromium found in natural whole wheat. Molasses has twenty-three times as much chromium as refined sugar, grape juice has forty times more, orange juice has thirty-four times as much, and maple syrup has five times as much. Brown rice has four times as much chromium as white rice.

So-called brown sugar is almost as depleted of chromium as white sugar. As stated before, brown sugar is usually made from refined white sugar with only a small amount of molasses added to darken the color. The same goes for most so-called raw sugar. Burnt sugar, called caramel on the label, is often added to refined white bread, misleading the buyer into thinking it contains more nutrients. If our foods were eaten without refining and processing, in a state close to the state in which they were grown, and harvested from fertile and composted soils, we would still be receiving adequate dietary chromium along with many other micronutrients that are largely deficient in our present diet.

Commercial fertilizers contain only nitrogen, potassium, and phosphorus, with the occasional addition of calcium in lime. Rarely are soils replenished with trace elements like chromium.

Populations that subsist on more naturally grown foods do not suffer epidemics of atheosclerosis, diabetes, heart attacks, senility, strokes, cancer, or arthritis. In many such cultures, people get smarter with age, rather than senile, and the eldest are looked to as sages and leaders of their societies.

You may ask, "Why not just take a chromium tablet each day to make up the difference?" That can be helpful, and, combined with many other trace elements, is recommended if taken in the proper form. But chromium, as you see it on the shiny bumper of an automobile, is not the same as the chromium we need in our diets. Many people cannot utilize even the chromium found in many lesser-quality supplements. And chromium in the wrong form or in excess can be toxic.

It's difficult to improve on naturally occurring nutrients as they occur in food. A healthy diet contains many undiscovered substances that cannot be supplemented in vitamin and mineral tablets.

The active form of chromium in the human body is a complex organic substance called glucose tolerance factor (GTF). GTF is a large molecule containing one chromium atom, with an electrical valence of +3, chemically bonded to two molecules of vitamin B-3 (niacin), and several amino acids. GTF can be obtained from whole foods and is necessary for proper insulin metabolism. Many people do make adequate GTF in their own bodies, when chromium is present, but most people cannot make enough and must obtain additional GTF from their food.

GTF has characteristics of a hormone. It is released by specialized tissues, such as the liver, into the bloodstream in tiny amounts in response to a meal containing sugar or starch. It therefore fits the definition of a hormone, just like insulin or adrenaline.

Because many people, especially older people, cannot manufacture enough GTF in their own bodies and must consume additional amounts in food, it also meets the definition of a vitamin. GTF contains the essential trace element chromium. Vitamin B-12 contains the trace element cobalt. Vitamins that have minerals as cofactors are not uncommon. The distinction between vitamins, minerals, and hormones can therefore become blurred.

A nutritional supplement shown in clinical studies to possess high GTF activity is patented under the trade name ChromeMate. Chromium picolinate and chromium polynicotinate also provide some GTF activity. The exact chemical structure of GTF is still uncertain.

To avoid problems of multiple nutritional deficiencies we should return to a more natural diet, a diet closer to that of our ancestors. For optimum intake, it is also necessary to take a broad-spectrum supplement on a daily basis (as described in chapter 15).

Sugar is hidden in many foods and beverages. Natural foods, as they grow on organically composted fertile soils, contain all of the nutrients necessary for proper health. It's only when we fractionate, process, refine, and tamper with those foods that the degenerative diseases of industrialized nations become epidemic.

Sophisticated laboratory testing shows that virtually all Americans are deficient or suboptimal, not just in chromium but in calcium, magnesium, manganese, zinc, copper, boron, vanadium, molybdenum, selenium, and many other elements.

What's the solution? Avoid, as much as practical, sugar and sugar-containing foods such as candies, cakes, pies, ice cream, soda pop, Kool-Aid, jellies, jams and so-called "fruit drinks." Avoid white flour products such as white bread, refined noodles, spaghetti, and biscuits. Make or buy foods made from 100 percent stone-ground, whole-grain flour whenever practical. Eat brown rice instead of white rice. Use honey or molasses in moderation, instead of refined sugar. Drink unsweetened natural fruit juices, also in moderation, or better yet, get your natural sugar from the whole fruit. If you eat canned fruits, use those packed in water or in their own juice, perhaps with a little honey but not in heavy syrup. Moderate use of fats and oils in the diet. Avoid fried and greasy foods.

If you don't want hardening of the arteries, heart attacks, strokes, diabetes, arthritis, low blood sugar, or senility in old age, the first step is to eat a proper diet.

AMA PUBLISHES PSEUDOSCIENCE TO DISCREDIT CHELATION

In January, 2002, the American Medical Association published a deceptively worded unscientific study alleging to disprove EDTA chelation as a treatment for heart disease—the so-called Calgary PATCH study.

That study seems carefully designed in an attempt to disprove chelation from the outset. Only one-fourth the number of patients needed for statistical significance was included. Patients most likely to benefit were selectively excluded. Most patients in the study had only minor symptoms, and thirty percent had no symptoms at all. It is not possible to study angina in patients who do not have angina. Twice as many patients who previously experienced myocardial infarctions were placed in the EDTA-treated group as were placed in the placebo group.

Approximately twice as many patients in the misrepresented "placebo" group were given potent anti-anginal drugs. That would obscure comparative improvement in EDTA-treated patients, who received only half as much anti-anginal drug therapy.

The exercise protocol was bizarre. The investigators failed to screen for reproducibility as a condition for entry. Accepted scientific guidelines were ignored. The primary endpoint was not clearly defined. The type of electrocardiographic ST-depression used as an endpoint is non-specific and no longer accepted as diagnostic for coronary disease.

Four patients in the placebo group and none in the EDTA-treated group underwent angioplasty during the one-year follow-up after chelation, and EDTA-treated patients showed better improvement in maximal oxygen consumption.

A detailed scientific analysis of this study can be found on the Internet at www.drcranton.com/sham.htm.

INDEX

medical commentary on, ix-xiii,
103-115
"naturalness" of, 30
New Zealand study, 126-128, 257-
258
oxygen toxicity, 277
paying for, 89-102
as a preventative, 24, 152
programmed cell death, 265
protection against oxygen free
radicals, 277-280
questions about, 151-162
research into, 4-5, 255-260
supposed dangers of, 31
and tobacco use, 159, 208-211,
285, 303
toxic heavy metals, 294-297
treatment and prevention of diseases of aging, 302-304
in treatment of atherosclerosis
and diseases of aging, 253-330
tumors, 301-302
workings of EDTA, 30-31
Chelation Therapy (Soffer), 6
chelators
industrial uses of, 30
oral or rectal, 147
Cheraskin, E., 6
chest pains, xx, 10-11, 45-46
cholesterol
metabolism, 50-51, 269-273
oxidized, 266, 286
cholesterol-lowering drugs, 172, 272
chromium, 356-365
chromosomes, 262
citric acid, early chelating compound, 2
Clarke, Norman E., Sr., 4
claudication, intermittent, 124-125
clinical research, 116-135
clot-dissolving enzymes, 172-173
coenzyme Q-10, 216, 267
cold fingers and toes, 46
collagen health, 76
commentaries, on chelation, ix-xiii,
103-135

competition with surgeries, 97
coronary artery angiograms, 47
coronary artery bypass grafting
(CABG), xx, 11
coronary artery disease, and precip-
itous rise in trace elements, 35
coronary artery spasms, 47-48
Coronary Artery Surgery Study
(CASS), 230-231
cost of chelation, 89-102
course of treatment
beginning, 151-152
interrupting, 155-156
length of, 158
cramps, leg, 45-46
Cranton, Elmer M., ix-xii, xv-xvi,
253
contacting, 330
Creutzfeldt-Jakob disease, 77
Cutler, Neal R., 78

d-penicillamine, 77
dairy products, 205
damage to arteries, causes of, 59
dangers
supposed, 31
what you haven't been warned
about, 136-147
Danish study, 124-126, 255-257
deferoxamine, 83
degenerative disease
devastating increase in, 81
free radical causes of, 265-266
and high blood calcium, 57
dementias, 77-83, 290
Demopoulos, Harry B., 86
depletion, of trace elements, 358-
365
deposits, calcium, 54-55
desserts and snacks, 205
diabetes, xxvi, 72-73, 158
diet drinks, 196, 202
dietary deficiencies, 50, 349-356
dietary fat, quality of, 283
dietary fiber, increasing, 196

ABOUT THE AUTHOR

 Elmer M. Cranton, M.D., graduated from Harvard Medical School in 1964. He served as president of Smyth County Medical Society, Virginia, and was chief-of-staff of a U.S. Public Health Service Hospital. He is a fellow and past president of the American College for Advancement in Medicine and a charter fellow of the American Academy of Family Practice. Dr. Cranton has been certified and recertified as a diplomate of the American Board of Family Practice. He served for many years as editor-in-chief of the *Journal of Advancement in Medicine* and has authored many books and scientific articles in the field of medicine. Dr. Cranton currently oversees two clinics at the addresses below.

Elmer M. Cranton, M.D.
Mount Rainier Clinic
503 First Street South, Suite 1
P.O. Box 5100
Yelm, WA 98597-5100
Phone: (360) 458-1061
FAX: (360) 458-1661
e-mail:
drcranton@drcranton.com
website: www.drcranton.com

Elmer M. Cranton, M.D., and
Eduardo Castro, M.D.
Mount Rogers Clinic
799 Ripshin Road
P.O. Box 44
Trout Dale, VA 24378-0044
Phone: (540) 677-3631
Fax: (540) 677-3843

RELATED TITLES

A Textbook on EDTA Chelation Therapy
Second Edition
Edited by Elmer Cranton, M.D.
With a foreword by Linus Pauling, Ph.D.

The most comprehensive technical source on EDTA chelation therapy available by the author of *Bypassing Bypass Surgery*, Elmer Cranton, M.D. Cranton presents the latest research on this effective, safe, and sensible alternative to bypass surgery in a fully revised and expanded edition.

Hardcover • 565 pages • ISBN 1-57174-253-0 • $75.00

Questions from the Heart
Answers to 100 Questions about Chelation Therapy, A Safe Alternative to Bypass Surgery
Terry Chappell, M.D.

If you're looking for an outstanding introduction to chelation therapy, *Questions from the Heart* is perfect for the first-time patient or a doctor's waiting room. In a Q&A format, this straightforward and simple book explains the basic concepts of chelation therapy in a way that everyone can easily understand.

Trade paper • 136 pages • ISBN 1-57174-026-0 • $10.95

Racketeering in Medicine
The Suppression of Alternatives
James P. Carter, M.D., Dr.P.H.

This daring book presents disturbing evidence of medical industry corruption. Dr. Carter's magnificent exposé of the ills of modern medicine and the greed of the corporations who control public health will open your eyes forever.

Trade paper • 392 pages • ISBN 1-878901-32-X • $12.95

For the most precious gift of all
— the gift of health —
send copies of *Bypassing Bypass Surgery*
to people dear to you.

Medex Publishers
P. O. Box 1000
Yelm, WA 98597-1000
Ph: 800-742-5682 or 360-458-1077 • Fax: 360-458-1661

Please send *Bypassing Bypass Surgery* to the address listed
below. I am enclosing **$22.45** ($17.95 plus $4.50 to cover
postage and handling). Credit card, check or money order.

Credit Card: ☐ VISA   ☐ MasterCard   ☐ Discover

Card
Number _____

Expiration date: _____

Signature_____

Name _____

Address _____

City_____State/Zip_____

Phone (_____) _____

NOTE: Should you wish to include a gift card,
enclose it with your order.
Call for volume discounts.

For the most precious gift of all
— the gift of health —
send copies of *Bypassing Bypass Surgery*
to people dear to you.

Medex Publishers
P. O. Box 1000
Yelm, WA 98597-1000
Ph: 800-742-5682 or 360-458-1077 • Fax: 360-458-1661

Please send *Bypassing Bypass Surgery* to the address listed
below. I am enclosing **$22.45** ($17.95 plus $4.50 to cover
postage and handling). Credit card, check or money order.

Credit Card: ☐ VISA   ☐ MasterCard   ☐ Discover

Card
Number _____

Expiration date: _____

Signature_____

Name _____

Address _____

City_____State/Zip_____

Phone (_____) _____

NOTE: Should you wish to include a gift card,
enclose it with your order.
Call for volume discounts.